THE AFRICAN TRYPANOSOMES

World Class Parasites

VOLUME 1

Volumes in the World Class Parasites book series are written for researchers, students and scholars who enjoy reading about excellent research on problems of global significance. Each volume focuses on a parasite, or group of parasites, that has a major impact on human health, or agricultural productivity, and against which we have no satisfactory defense. The volumes are intended to supplement more formal texts that cover taxonomy, life cycles, morphology, vector distribution, symptoms and treatment. They integrate vector, pathogen and host biology and celebrate the diversity of approach that comprises modern parasitological research.

Series Editors
Samuel J. Black, *University of Massachusetts, Amherst, MA, U.S.A.*
J. Richard Seed, *University of North Carolina, Chapel Hill, NC, U.S.A.*

THE AFRICAN TRYPANOSOMES

edited by

Samuel J. Black
University of Massachusetts, Amherst, MA

and

J. Richard Seed
University of North Carolina, Chapel Hill, NC

KLUWER ACADEMIC PUBLISHERS
Boston / Dordrecht / London

Distributors for North, Central and South America:
Kluwer Academic Publishers
101 Philip Drive
Assinippi Park
Norwell, Massachusetts 02061 USA
Telephone (781) 871-6600
Fax (781) 681-9045
E-Mail < kluwer@wkap.com >

Distributors for all other countries:
Kluwer Academic Publishers Group
Distribution Centre
Post Office Box 322
3300 AH Dordrecht, THE NETHERLANDS
Telephone 31 78 6392 392
Fax 31 78 6546 474
E-Mail < services@wkap.nl >

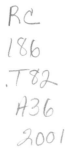

RC
186
.T82
A36
2001

 Electronic Services < http://www.wkap.nl >

Library of Congress Cataloging-in-Publication Data

The African typanosomes / edited by Samuel J. Black and J. Richard Seed.
    p. cm. – (World class parasites ; v.1)
   Includes bibliographical references and index.
   ISBN 0-7923-7512-2 (alk. paper)
    1. African typanosomiasis. I. Black, Samuel J., 1947– II. Seed, J. Richard, 1937-III.
Series.

RC186.T82 A36 2001
616.9'363—dc21

                                                                        2001038475

Printed on acid-free paper.
Printed in the United States of America

The Publisher offers discounts on this book for course use and bulk purchases. For further information, send email to <joanne.tracy@wkap.com> .

TABLE OF CONTENTS

PREFACE

African trypanosomes are flagellated protozoa that inhabit the extracellular compartment of host blood, in the face of the immune system which they flout by switching among distinct antigenic types. The spindle-shaped parasites are about 20 μm long and about 2 μm in diameter at their widest point, have a single flagellum and are motile. The African trypanosomes are transmitted to mammals in saliva deposited by biting tsetse flies (*Glossina spp*) in which they undergo cyclic development. Tsetse infest the humid and semi-humid zones of sub-Saharan Africa. There are several species of Glossina that are capable of transmitting trypanosomes, and also several species of African trypanosomes, some of which have a profound effect on the economy of sub-Saharan Africa. Among these, *Trypanosoma brucei brucei*, *T. congolense* and *T. vivax* are notable because they cause Nagana, a wasting and generally fatal disease in cattle, while *T. brucei rhodesiense* and *T. b. gambiense* are notable because they cause fatal sleeping sickness in people.

Nagana is endemic throughout the humid and semi-humid zones of Africa coincident with the distribution of tsetse, which infest an area of some 10 million km^2 embracing 36 countries. About 30% of Africa's cattle graze on the fringe of the tsetse habitat, many sustained by chemotherapy and tsetse control programs. Regions of high tsetse and trypanosome challenge, which account for some 70% of the humid and semi-humid zones of sub-Saharan Africa, are devoid of cattle and hence of integrated cattle and market garden systems. The absence of cattle and other domesticated animals for traction, forage conversion to fertilizer and livestock contribution to the small holder economy, makes the classic agrarian model of societal development inapplicable to this region.

Some species of trypanosomes have spread beyond Africa. *T. evansi* has been selected through evolution to be completely independent of tsetse-transmission, is spread through much of Asia and the middle East and causes debilitating disease in water buffalo. This can be a severe economic blow to small holder farmers who use these animals for traction and forage conversion. *T. vivax* has spread from Africa to parts of S. America where it is mechanically transmitted among cattle by biting flies causing a loss of condition and productivity. These biting fly-transmitted trypanosomes are of considerable economic significance but do not strangle economic development to the same extent as their tsetse-transmitted cousins in sub-Saharan Africa.

Human trypanosomiasis is restricted to sub-Saharan Africa and has a narrower distribution than nagana. It has been curbed historically by vector avoidance and control programs, combined with disease surveillance and treatment. However, these curbs have been eroded by population pressure, poverty, irrational politics and war. Sleeping sickness has reached epidemic proportions in Angola, the Democratic republic of the Congo, parts of Uganda and Sudan. Furthermore, its prevalence is high and increasing in Cameroon, Cote D'Ivoire, Guinea, Mozambique, Tanzania and Chad. Accurate figures for disease prevalence are available for only about 10% of the affected region but based on these figures it is estimated that some 300 to 500 thousands of people are infected. These are staggering numbers given that sleeping sickness is invariable fatal if untreated, the required treatment can be prolonged and expensive, and cured individuals remain susceptible to re-infection.

Many forces are re-modeling the face of modern Africa. Political upheaval linked to intertribal strife is re-distributing the human population and in many instances dismantling public health measures, increasing disease exposure and preventing diagnosis and treatment. Diseases that were once thought to be controlled, namely, malaria and trypanosomiasis, have flared to epidemic proportions and the need to control new diseases, e.g., AIDS, vies with the old for scarce economic resources. There is no question of singling out one or other disease as deserving of special attention. All of the diseases, as well as the breakdown in public health and political systems, which have facilitated their spread, are deserving of the most focused and urgent attention.

What contribution can scientists make to the containment and control of parasitic diseases such as trypanosomiasis? We can: (i) accurately describe the field situation, (ii) advocate the effective use of currently available strategies of parasite and vector control, (iii) elucidate new cause and effect relationships thus allowing the development of new methods of vector and parasite control, (iv) improve communication among scientists and (v) improve communication between scientists and administrative bodies that allocate resources, in order to gain additional support for parasitological research and its implementation. The essays in this volume address all of these activities. Readers will find that our contributors occasionally disagree about the solutions that will be most effective to control trypanosomiasis and even on where the greatest focus of attention and economic support should be placed.

Diversity of opinion is, we believe, good and the willingness of investigators to express their opinions hopefully helps the reader to appreciate the complexity of our enterprise to defeat the African trypanosomes.

Samuel J. Black , John R. Seed, 21st Feb 2001

AFRICAN TRYPANOSOMIASIS: FAILURE OF SCIENCE AND PUBLIC HEALTH

D.H. Molyneux
Liverpool School of Tropical Medicine,Pembroke Place
Liverpool L3 5QA

ABSTRACT

This essay contrasts the on-going scientific effort to characterize trypanosome biochemistry and molecular biology and to resolve the host: parasite relationship, with the pressing need to apply known effective technologies to control of trypanosomiasis and tsetse flies in the field.

Keywords disease control, scientific curiosity vs applied science

The discovery of the association of trypanosomes with the disease of humans and cattle spans a century when some of the greatest changes in human history have occurred, driven by science, technology and culture. During this period trypanosomes and their vectors have been the subject of intense study. Today we know more about the parasites and their tsetse fly vectors than most eukaryotic cells and insects respectively. However, the public and animal health problems remain. Students are commended to read the great historical accounts of trypanosomiases and tsetse flies for fundamental information (Laveran & Mesnil, 1907; Buxton, 1955; Mulligan, 1970; Ford, 1971; Hoare, 1972; Lumsden & Evans, 1976 & 1979; Stephens, 1986; , Leak, 1998; Hide et al., 1997). This essay presents a personal view of the field as it presently stands.

Characteristically and ironically the title of a recent text (Hide et al., 1997) is trypanosomiasis and leishmaniasis: Biology and Control; of 22 chapters in the book, only 5 deal with public health aspects of the diseases, socio-economic impact, control strategies, chemotherapy and impact of trypanosomes in reproduction of small ruminants. The remaining 17 chapters are devoted to biochemistry, cell and molecular biology. This and similar volumes have led me to the thesis proposed in the title of my essay and laid out below. Current students should read particularly reports of the Sleeping Sickness Bulletins from Bruce and colleagues in Uganda and the studies of Thomas, Dutton and Breinl on the discovery of Atoxyl as a means of treating sleeping sickness (Thomas, 1905) as well as the work of the Portuguese in Principe where eradication was achieved

by the use of sticky material on plantation workers backs
(Maldonaldo, 1910). As we approach the centenary of these
observations it is perhaps appropriate to reflect that Melarsoprol and
hence arsenic (notwithstanding the potential of
Difluoromethylornithine (DFMO)) remain the basis of treatment for
late stage human trypanosomiasis   Pepin et al (2000) have recently
published the outcome of a multicentre randomised controlled trial of
short course eflornithine in Gambiense sleeping sickness; they
compared a 7 day with a 14 day course (100mg/kg every 6 hours
intravenously). There were 321 patients randomised in four centres
(Congo, Cote d'Ivoire, Democratic Republic of Congo and Uganda)
of which 274 were new cases and 47 relapses. The study concluded
that the 7 day course was appropriate for relapse cases but that for
new cases the 7 day course was inferior to a 14 day regimen. Van
Nieuwenhove (2000) concludes "Thus melarsoprol will remain the
first line drug for the foreseeable future".

Tsetse control for human disease control is rarely
implemented within existing health systems and probably will never
be unless supported by donors. Today sleeping sickness remains one
of the most serious of the disease of Central and Equatorial Africa
(Ekwanzala et al., 1996) untouched by the publicity of Ebola virus,
HIV, malaria and TB but having an insidious depopulating effect on
huge areas of Angola, Congo, South Sudan, Central African Republic,
Northern Uganda and Congo Brazzaville – all countries characterised
by social and civil unrest (Van Nieuwenhove, 2000). In animal
trypanosomiasis, because the disease does not affect
western/developed livestock, no new drug has entered the market for
30 years.

The historic campaigns against sleeping sickness by the
former colonial powers actually succeeded in controlling the disease –
classical surveillance (however irregular in terms of area/or
population coverage), diagnosis (by gland puncture or blood film
however insensitive) and treatment (however unpopular and with side
effects) controlled the disease throughout Central and West Africa
supplemented in Francophone Africa by compulsory 6 months
pentamidine (lomidine) prophylaxis as a compulsory 6 monthly
intervention and in parts of English speaking Africa by tsetse control.
These were "vertical" programmes. Vertical today being a
"derogatory" health system terminology for an intervention which is
successful but probably not deemed cost-effective, sustainable or
capable of fitting into a generalised concept of health care
management or of decentralisation and devolution to district level

decision making and planning. The problems of African trypanosomiasis contrast with its sister vector borne disease – in South America – Chagas' disease caused by *T. cruzi*. The latter has been successfully controlled by house spraying against *Triatoma infestans* as a result of national government commitment to a long term programme with standardised evaluation and monitoring, succeeding after 15 years in achieving the certification of freedom from transmission from *T. cruzi* in Chile, Uruguay and most of Brazil. In onchocerciasis endemic areas of 11 countries in West Africa blindness no longer is a public health problem as a result of vector control by larviciding or donated drug distribution (Mectizan). The reason for these successes have been articulated by Molyneux & Morel (1998) – donor and country commitment, clear public health objectives, operational research, adaptable strategies, quality staff, transparent management, acceptable evaluation and monitoring, strong advocacy. These successes contrast spectacularly with the failures in sleeping sickness.

Sleeping sickness as a disease is an intractable problem when national health services are in decline, when the human resource capacity is limited, and donor funds are restricted and focused on the big three – malaria, tuberculosis (TB) and human immuno-deficiency virus (HIV). The disease is not "on the screen" in public health terms but many scientists have embarked on careers to study trypanosomes (not sleeping sickness note) in an attempt to develop a drug by rational biochemistry and molecular biology. The funding of studies on trypanosomes by a review board or study panel is somehow synonymous with sleeping sickness and hence international health and tropical disease control. Donors assuage their guilt by funding such studies with only a marginal, if any, impact on public health. Studies on the biochemistry, immunology and molecular biology of African trypanosomes have yielded probably one output of relevance. It is the only outcome from the studies of antigenic variation – the Card Agglutination Test for trypanosomiasis (CATT), which can be identified as having been used in sleeping sickness control. However, although this low technology test is appropriate for the field diagnosis of trypanosomiasis, many countries cannot afford to utilise it. In fact, only donor support in restricted situations has seen it deployed.

The elegant trypanosome is easily cultured now and grown in laboratory animals and with excellent primate models. These technical advances have yielded an excess of scientific papers but no new drug. There is probably more information about the biochemistry and molecular biology of trypanosomes than any other non-mammalian cell type and a great deal is known about the differences between trypanosomes and mammalian cells, but there is no new therapeutic product. Let us remember the famous introductory paragraphs of every grant application we have read, in which the proposed studies are predicted to lead to novel strategies of trypanosomiasis control, or novel therapeutic agents. Why will scientists not admit that they are simply fascinated by the organism and admit that the studies they propose won't have any public health impact?

The optimistic launch of UNDP/World Bank/ WHO Special Programme for Research and Training in Tropical Diseases (TDR) in the 1970 was paralleled by the creation of the International Laboratory for Research in Animal Diseases (ILRAD) in Nairobi with a mandate to develop a vaccine for animal trypanosomiasis. Given the promise of the new biology, nothing seemed impossible; the Board of ILRAD largely drawn from the north with great immunological expertise proclaimed with optimism that a vaccine was achievable (for all three animal trypanosome species!). ILRAD's mandate did not unfortunately recognise the value of vector control and ignored the seminal ecological studies of Vale and colleagues as well as French entomologists which resulted in the development of significant cost-effective interventions against tsetse for a fraction of the cost of ILRAD's annual budget (Vale, 1974; Vale & Hargrove, 1975).

Those with long memories will also remember the publicity received by the announcement of the cultivation of *T. brucei* in vitro (Hirumi et al., 1977) an excellent piece of science but erroneously proclaimed as the key breakthrough in vaccine development. Regrettably the critics of ILRAD, and there were many, had their voices suppressed. The bench scientists did not appreciate the field situation. Research was driven by expertise out-of-step with field needs which were exacerbated by a decline in animal health systems and services generated by structural adjustment policies and a continuous decline in national resources entering the public sector.

Sleeping sickness research was supported conspicuously well by WHO/TDR from the late 1970s; research on diagnosis, epidemiology, and entomology resulted in progress; host parasite relationship in vectors, pathology and drug targeted biochemical research also developed. Refinements in diagnosis emerged – CATT, mini-anion exchange centrifugation; identification by new methodologies of animal reservoirs; refinements of tsetse traps and development of impregnated targets; better understanding of pathology from laboratory studies and standardised post-mortem studies. However, the lack of capacity to sustain improved interventions (aside from chemotherapy which remained melarsoprol based despite the emergence of DFMO and potential of Niafurtimox for relapse cases) and the break-up of civil order in key endemic countries were the initial determinants of the current problems. The sleeping sickness community remained resilient to communication with health systems researchers who take a macroscale view of how disease control options might be integrated into declining impoverished systems. In addition there has been no global advocate of stature for sleeping sickness, probably because it is not amenable to a magic bullet and hence is not sexy enough to attract a high profile advocate (like onchocerciasis or Guinea worm) nor does it not threaten western democracies (like HIV) and TB. The need remains for a planned assessment of what capacity exists and where – be it Non-Governmental Development Organisations (NGDO) (never seriously recruited into sleeping sickness control), missions, district hospitals in endemic areas, existing strong programmes in other diseases (e.g. Guinea worm, onchocerciasis) training health workers at the periphery just to recognise some key symptoms to alert District Medical Officers (DMO) or anyone to the potential problems.

Sleeping sickness has not been treated as an emergency disease. However, compared to Ebola it is a classic re-emergent, but chronic, infection in many parts of Africa. Ebola kills 70%+ of its victims quickly; sleeping sickness kills them all slowly. Advocating the emergence of sleeping sickness has not been seriously taken up or more importantly sustained with the exception of CARE and Medécin san Frontieres (MSF). Donor support has virtually waned as HIV, TB and malaria have claimed the international agenda within the G8 and the Hashimoto initiative which has embraced helminth diseases. Trypanosomiasis in Africa and leishmaniasis, for not dissimilar reasons, remain the Cinderallas in public health terms despite lavish expenditures by the science community on the intricacy and fascination of the organisms (Van Nieuwenhove, 2000).

Tsetse research and control embraces a fascinating history; an insect which is the most fragile in terms of its apparent biological inefficiency. A female producing one offspring every 9-12 days and both sexes depending on blood meals – hardly at first sight a recipe for success as well as an inefficient vector of *T. brucei*. *Glossina* vulnerability to control by various means is not in doubt – Maldonado (1910) described eradication on Principe; hugely successful trapping in Zululand in the 1920s; game destruction and bush clearing in the

1940s and 50s (notwithstanding ethical and ecological concerns) and subsequently DDT based ground spraying of nesting sites which made a huge impact in N.E. Nigeria, Uganda and Zimbabwe during 1950s and 60s until the ecological concerns of chlorinated hydrocarbon use were voiced justifiably but often with little regard for the proven public and animal health benefits of DDT. The work on traps undertaken by Southern African workers in the 1920/30s and Morris & Morris (1949) and Morris (1950) in Ghana as well as bush clearing programmes in 1938-44 (Morris, 1950) which controlled sleeping sickness had been largely, if not totally, ignored until the 1970s.

Impregnated traps and targets have also made way for the approach of using the host itself as the target and using pour-ons formulated to have long persistence, be non-toxic and spread through the animal's coat following a simple application. Again the first use of pour-ons date back to 1940s/50s. Whiteside (1949) used DDT in groundnut oil against a population of *G. pallidipes* as well as in the 1950's after *G. palpalis* returned to the island (quoted by Jordan, 1986). Pour-ons of deltamethrin can now be used as an effective intervention controlling both tsetse and ticks (Bauer et al., 1995).

Coincidentally Vale (1974) and Challier & Laveisierre (1973) resurrected the trapping studies independently in Zimbabwe and Burkina Faso. Taking different but equally valid approaches committed biologists produced useable cost-effective tools. Vale and Hargrove's elegant studies on attractants, behaviour, ecology and target design supplemented by other studies on urinary phenols as attractants by the International Centre for Insect Physiology and Ecology (ICIPE) lead to deployment in many sites of impregnated targets which incorporated odour bait (acetone, octenol, and cattle urine) for *G. morsitans* group flies (Vale & Hargrove, 1975). A cheap sustainable viable technology was available. For *G. palpalis* group flies the biconical trap (Challier & Laveissiere, 1973) evolved into different models for different settings and in field trials demonstrated a remarkable efficacy in reducing the tsetse fly population by over 99% when impregnated with deltramethrin. Lancien (1991) applied such traps to assist in the control of the Uganda sleeping sickness epidemic in the 1980s allied to treatment programmes of those infected. Over the same period refinements in the use of aerial spraying using ultra low volume (ULV) spray from fixed wing aircraft were effective on a large scale against animal disease vectors in Botswana and helicopters (expensive and logistically complex) were deployed for residual application against riverine flies in West Africa (Molyneux et al., 1978 and Speilberger et al., 1979).

The sterile male release approach was in the 1970s onwards widely advocated by the US Agency for International Development (USAID) and the International Atomic Energy Agency (IAEA) as the ultimate solution to tsetse control despite being irrelevant for sleeping sickness control. The naivety of this approach remains a mystery to the author who has the view that

1. more than one species of *Glossina* are usually present in an area
2. it takes around 2-3 years to build up colonies to sufficient numbers to release sterile males;
3. Insecticides (ULV spray or traps) need to be used to reduce populations by 90% before release. Two ULV sprays are usually necessary. Why not undertake three more cycles to remove the whole fly populations? This could of course commence two-three years earlier and save capital and recurrent costs of colonies which seem unsustainable for national livestock services. Shipment of flies from Vienna to Nigeria never seemed a feasible option!
4. Success against *G. austeni* in Zanzibar was proclaimed as a huge breakthrough. Zanzibar is not mainland Africa! One species – a small isolated island population. Meanwhile, we have traps and ULV insecticide techniques, which are perfectly adequate, traps being rapidly deployed, are cheap and more sustainable by local communities.

The establishment of the Partnership against African Trypanosomiasis (PAAT) between international agencies would, one hopes, have injected realism into the debate as to what is feasible. However, a recent newsletter proclaims that African Heads of State have signed up to the eradication of tsetse flies (PAAT, 2000). The proponents of this expectation should examine the current resourcing of the health and agricultural sectors in Africa, the competing needs in the health sector of TB, HIV and malaria, the availability of effective donated tools for other diseases and their own record over the last decades in mobilising resources for trypanosomiasis. The proposal is driven again by the SIT lobby. The prospects of eradicating forest group flies (*G. fusca*) or the *G. palpalis* in Central Africa using the SIT approach demands that huge colonies are *Glossina* established for each species. We are expected to endorse this in areas of many countries where civil unrest persists and access to any health facility or animal health advice is non-existent. Such suggestions bear no regard to financial or logistic reality. It is regrettable that individuals and organisations are in the process of compounding earlier failures to control the diseases by even suggesting *Glossina* can be eradicated. Here is an example of lessons never being learned. The costs of eradicating

G. austeni in Zanzibar was US$85 million. One can use this value to estimate the cost of "eradicating" the remaining *Glossina* in an area of 10.5 million km^2 infected by 23 species. The area of Zanzibar is 1,700km^2 hence a single calculation suggests that the total cost of eradication in Africa would be 23 x 85 million x 9.9^6 (10 million km2 minus area infected in Zanzibar 1,700 km^2) i.e. costs of Africa eradication in 10^7 km^2 using Zanzibar costs are $1.9 x 10^{16} US.

The parallels between the fixation with sterile male release and the need to pursue trypanosome biochemistry at the expense of the public and animal health aspects reflect an imbalance in investment towards what might make an impact. In the field, the need to use appropriate technologies, strengthen local capacity at an appropriate level and remember at least in tsetse control the wheel has been invented several times.

Whilst investment in the science of trypanosomes will continue our track record in producing designer drugs, is abysmal, in fact non-existent. No new product used in human tropical diseases has been developed through any rational process. The products for helminth diseases derive from the veterinary market. If animal trypanosomiasis was a disease of western livestock we might have a safe human trypanoside. As the late Jim Williamson said there has been far more reviews of trypanosome chemotherapy than usable products.

The solution now to a huge public health problem surely lies in ensuring we use what we have. Engage local health workers, ensure they are adequately trained and seek to develop sustained advocacy for funds to make an impact; learn lessons from other programmes and refocus on the practicalities of public and animal health, seek solutions at the field level whilst science is given a final opportunity to solve what ought to be a simple problem given the major differences between trypanosomes and mammalian cells. Meanwhile we should deploy diagnostic and vector control tools assiduously and cost-effectively within sustainable health systems, advocate for disease control as opposed to rational drug design or vaccinology and insist that sleeping sickness, as an emerging or re-emergent disease, is as important as Ebola or HIV in many Central Africa communities at present condemned.

REFERENCES

Bauer, B., S. Amster-Delafosse, P.H. Clausen, I. Kabore and J. Petrich-Bauer. 1995. Successful application of a deltamethrin pour-on to cattle in a campaign against tsetse flies (*Glossina* spp) in a pastoral zone of Samorogonan, Burkina Faso. Tropical Medicine and Parasitology. 46, 183-9.

Buxton, P.A. 1955. The natural history of tsetse flies. London School of Hygiene and Tropical Medicine. Memoir 10, 816 pages, H.K. Lewis, London.

Challier, A. & C. Laveissiére. 1973. Un nouveau piege pour la capture des glossines (*Glossina*

Diptera, Muscidaed): description et essais sur le terrain. Cali ORSTOM ser Ent. Med. Parsite. 11, 251-62.

Ekwanzala, M., J. Pepin, N. Khonde, S. Molisho, H. Brunel, P. de Wals. 1996. In the heart of darkness: sleeping sickness in Zaire. Lancet 348, 1427-30

Ford, J. 1971. The role of the trypanosomiases in African ecology, 568 pages, Clarendon, Oxford.

Hide, G., J.C. Mottram, G.H. Coombs and P.H. Holmes. 1997. Trypanosomiasis and leishmaniasis. Biology and control. 366p. CAB International, Oxford.

Hirumi, H, Doyle, J.J. & Hirumi, K. 1977. African trypanosomes: cultivation of animal infective *T. brucei* in vitro. Science, 196, 992-994.

Hoare, C.A. 1972. The trypanosomes of mammals. A zoological monograph, 749p, Blackwell, Oxford.

Jordan, A.M. 1986. Trypanosomiasis control and African rural development. 357p, Longman Harlow.

Lancien, J. 1991. Lutte contre la maladie du sommeil dans le sud est de l'Uganda par le piegage des glossines. Annales de Sociologie belge Med trop. 71, 35-47.

Laveran, A. and F. Mesnil. 1907. Trypanosomes and trypanosomiases translated by D. Nabarro, 538 pages, London Bailliere, Tindall & Cox.

Leak, S.G.A. 1998. Tsetse biology and ecology. Their role in the epidemiology and control of trypanosomiasis. CABI with ILRI, 568 pages, Wallingford Oxford, New York

Lumsden, W.H.R. and D.A. Evans (eds) 1976. Biology of kinetoplastids. Vol.I. 563 pages. Academic Press, London, New York and San Francisco.

Maldonado, B. 1910. (English abstract of Portuguese texts of 1906 and 1909) Sleeping sickness. Bureau Bulletin. 2, 26.

Molyneux, D.H. & C. Morel 1998. Onchocerciasis and Chagas disease control: the evolution of control via applied research through changing development scenarios. British Medical Bulletin, 54, 327-339.

Molyneux, D.H., D.A.T. Baldry, P. De Raadt, C.W. Lee & J. Hamon. 1978. Helicopter application of insecticides for the control of riverine *Glossina* vectors of African borne trypanosomiasis in the moist savanna zones. Annales de Societe belge Med. Trop. 58, 185-203

Morris, M.G. 1950. The persistence of toxicity in DDT-impregnated hessian and its use on tsetse traps. Bulletin of Entomological Research. 41, 259-88

Morris, K.R.S, & M.G. Morris. 1949. The use of traps against tsetse n West Africa. Bulletin of Entomological Research. 39, 491-528.

Mulligan, H.W. (ed). 1970. The African trypanosomiases. 950p. George Allen and Unwin. Ministry of Overseas Development, London.

PAAT Neswletter 2000. Chairman's report. No.7, October.

Pepin, S., Khonde, N., Maiso, F., Doua, F., Jaffar, S., Ngampo, S., Mpia, B., Mbulamberi, D., Kuzoe, F. 2000. Short-course eflornithine in Gambian trypanosomiasis: a multicentre randomized controlled trial. Bulletin of the World Health Organization, 78, 11, 1284-1295

Speilberger, U., B.K. Naisa, K. Hoch, A. Manno, P.R. Skidman & H.H. Coutts 1979 Field trials with the synthetic pyrethroids, permethrin, cyperpermethrin and decamethrin against Glossina (Diptera: Glossinidae) in Nigeria. Bulletin of Entomological Research. 69, 667-89

Stephen, L.E. 1986. Trypanosomiasis – a veterinary perspective, 551 pages. Pergamon Press, Oxford.

Thomas, W. 1905. Some experiments on the treatment of trypanosomiasis. British Medical Journal, 1140-2.

Vale, G.A. 1974. The responses of tsetse flies (Diptera: Glossindae) to mobile and stationary bait. Bulletin of Entomological Research, 67, 545-88.

Vale, G.A. & J.W. Hargrove . 1975. Field attractions of tsetse flies (Diptera Glossinidae) to ox odour: to effects of dose. Bulletin of Entomological Research. 56, 46-50.

Van Nieuwenhove, S. 2000. Gambiense sleeping sickness: re-emerging and soon untreatable?

Molyneux.

Bulletin of the World Health Organization, 78, 11, 1283
Whiteside, E.F. 1949. Experimental control of tsetse with DDT-treated oxen. Bulletin of Entomological Research, 40, 123-34.

THE PROGRAMME AGAINST AFRICAN TRYPANOSOMIASIS INFORMATION SYSTEM (PAATIS).

M. Gilbert [1], C. Jenner [2 3] J. Pender [2], D. Rogers[4], J. Slingenbergh [3], and W. Wint [5]

[1] Laboratoire de Biologie animale et cellulaire, CP160/12, Free University of Brussels, av. F.D. Roosevelt 50, B-1050 Brussels, Belgium
[2] Natural Resources Institute, Central Avenue, Chatham Maritime, Kent ME4 4TB UK
[3] Food and Agriculture Organisation, Via delle Terme di Caracalla, 00100, Rome
[4] Trypanosomiasis and Land Use in Africa Research Group, Dept of Zoology South Parks Road Oxford OX1 3PS UK
[5] Environmental Research Group Oxford Ltd, Dept of Zoology South Parks Road Oxford OX1 3PS UK – Author for correspondence.

ABSTRACT

The Programme against African Trypanosomiasis (PAAT) was established in 1995 and is managed by a joint secretariat composed of FAO, OAU/IBAR, IAEA and WHO. The programme aims to provide direction and focus to the control of trypanosomiasis within the broader context of food security, human health, rural development and sustainable agriculture. This paper describes the PAAT-Information System (PAATIS) which aims to assist decision-support for prioritisation of tsetse control areas and control strategy, thereby facilitating direct contact between programmes and providing a common source of information on tsetse and trypanosomiasis. The system is aimed at a broad range of users and will be available as a beta on a packaged CD-ROM. It will then be widely distributed for evaluation and to generate feedback. The final version will be upgraded on a regular basis and will be complimented by the activities of the PAAT email forum and the PAAT website.

Keywords Tsetse, Trypanosomosis, GIS, Decision Support, Control, Remote sensing, cattle, agriculture, cropping, impact

INTRODUCTION

The Programme against African Trypanosomiasis (PAAT) was established in 1995 and is managed by a joint secretariat composed of

FAO, OAU/IBAR, IAEA and WHO. The programme aims to provide direction and focus to the control of trypanosomiasis within the broader context of food security, human health, rural development and sustainable agriculture.

Development of the PAAT Information System (PAATIS) has been driven by the need for decision support at a continental and regional level to guide strategic decisions on tsetse and trypanosomiasis control in sub-Saharan Africa. This paper reviews the objectives, design, contents and outputs of the system and explores options for future development.

The development of the PAATIS was based in FAO, Rome supported by technical inputs from: the Environmental Research Group Oxford Ltd. (ERGO), the Trypanosomiasis and Land Use in Africa (TALA) Research Group, Department of Zoology, Oxford University and the Natural Resources Institute (NRI), Chatham, UK. Members from African Government Departments, Tsetse Control Divisions, Universities, NGOs, Researchers, Consultants and Donor Organisations were consulted at the design stage as were a wide range of stakeholders attending a variety of international seminars, workshops and meetings.

OBJECTIVES AND DESIGN

Decisions on system design and requirements evolved from the following objectives: to identify areas where the impact of tsetse and trypanosomiasis on agriculture is greatest; to provide decision-support on control strategy and support decisions made at national level; to improve co-ordination between organisations involved in control of tsetse and trypanosomiasis; and to provide data for research & development. The Information System is made up of three interacting components. The Geographical Information System (GIS) providing the capability for storage, display and analysis of layers of spatial data; the Resource Inventory containing country level tsetse and trypanosomiasis information; and the Knowledge Base allowing the user to query an extensive database of accepted literature. Complementary to the Information System, has been the formation and moderation of the PAAT Link (PAAT-L) email forum. With over 250 subscribers, this has evolved into a focal point for the release of general announcements and scientific debate. There is also a web site, hosted by FAO, which provides digests of the data held in the Information System, as well as details of the PAAT program and its activities (http://www.fao.org/paat/html/home.htm)

The system requires the user to have Arcview and Spatial Analyst installed and can be downloaded from *ftp://ergodd.zoo.ox.ac.uk/paatiszips* (view the readme txt). However, all components will be distributed on a CD to a range of stakeholders.

THE GIS

The GIS component of the PAATIS allows display and detailed analysis of geographical information related to African trypanosomiasis, providing an integrated common platform for continental scale data. Data available in PAATIS (Table 1) were compiled from a wide variety of sources as detailed below. Raster data have a 0.05 degree resolution.

Table 1: Data currently available in PAATIS Beta (v 1.0).

Data Description	Predicted	Source
Layers (Vector Polygons)		
Continental Regions	No	PAAT/ERGO
Countries	No	UNEP/GRID, Nairobi
Administrative Level 1 and 2	No	UNEP/GRID, Nairobi
Major Rivers	No	NRI
River Basins	No	USGS/EROS Hydro1K
Major Roads	No	NRI
Major Towns and Cities	No	NRI
National Parks	No	PAAT/ERGO
Forest Types	No	FAO
Lakes	No	PAAT/ERGO
Tsetse fly Distributions	No	ERGO/TALA/ILRI/FAO
Background (Images)		
Annual Rainfall *[†]	No	Cramer and Leemans
Human Population *[†]	No	Various – See text
Elevation *[†]	No	USGS
No Tsetse Species	No	ERGO/TALA/ILRI/FAO
Morsitans group	Yes	ERGO/TALA/ILRI/FAO
Fusca group	Yes	ERGO/TALA/ILRI/FAO
Palpalis group	Yes	ERGO/TALA/ILRI/FAO
NDVI *	No	TALA
Observed Cattle *	No	Various – See text
Observed Cultivation *	No	Various – See text
Predicted Cattle *[†]	Yes	PAAT/TALA/ERGO
Predicted Cultivation *[†]	Yes	PAAT/TALA/ERGO
Length of Growing Period *[†]	Yes	FAO/AGL/ERGO/TALA
Mammal Biodiversity	Yes	PAAT/TALA/ERGO
Farming Systems	Yes	PAAT/TALA/ERGO
Ecozones	Yes	PAAT/TALA/ERGO
Data Reliability, Cattle	N/A	PAAT/ERGO
Data Reliability, Cultivation	N/A	PAAT/ERGO

*Raster data layers marked * are accessible for local statistical calculations and those marked [†] are available for statistics tabulated by Country and/or Ecozone and/or Farming Systems (Mean, Max, Mean and Total when applicable).*

The tsetse distribution maps have been derived from those produced by Ford and Katondo in 1977, using remotely sensed satellite imagery of climatic indicators (such as temperature, rainfall and vegetation cover) to provide predicted distributions of the three species groups (*Morsitans, Palpalis* and *Fusca)*. A background layer of the number of tsetse species present is also provided. Details of the prediction methodology can be obtained from FAO (2000) or *ergodd.zoo.ox.ac.uk/tseweb/index.htm.*

The human population data given is combined from three sources: a global coverage of population number per image pixel from University of California at Berkeley provided by FAO; a population density coverage at the same resolution from the Consortium for International Earth Science Information Network (CIESIN: *www.ciesin.org*), derived from data collated by the National Centre for Geographic Information and Analysis (NCGIA: *www.ncgia.ucsb.edu*); and data from the Intergovernmental Authority on Drought and Development (IGADD) countries.

Two sets of cattle density and cropping percentage data are provided: 'observed' and 'predicted' (Figure 1). The former represents the national and sub-national census data covering the period between 1985 and 1999, available from a wide range of sources. For cattle these include: the International Livestock Research Institute; ERGO aerial survey archives and Government Agricultural Census data. The observed cropping data were obtained from: the Africa Data Dissemination Service; FAO AGDAT as used in the FAO GEOWEB service, produced by FAO GIEWS *(geoweb.fao.org/)*; ERGO/TALA aerial survey archives; transcribed Government Census data; and FAO GIEWS reports.

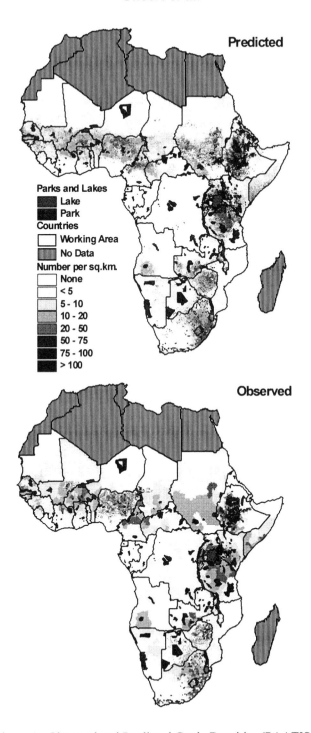

Figure 1: Observed and Predicted Cattle Densities (PAATIS, 2000)

This 'observed' information is largely at the level of administrative units, some of which are very large (Figure 1, bottom). This resolution has been increased by using stepwise multiple regression to establish statistical relationships between these observed data and a range of predictor variables including: satellite imagery, provided by the TALA Research Group, related to rainfall, temperature, vapour pressure deficit, vegetation cover and elevation; potential evapotranspiration; length of growing period; human population; and the number of tsetse species present. These data were extracted for some 12,000 sample points covering sub-Saharan Africa, and a separate relationship established for each of a number of ecozones (see below) occurring within each country. The resulting equations were then applied to the high resolution imagery to provide predicted maps of cattle and cultivation at a resolution of 5 kilometres. Full details can be found in FAO (1996c, 1997, 1998 &1999). All the predictive equations used were formally significant to at least the one percent level (p<0.01), and most substantially more so. The predicted distributions can thus be taken to be statistically acceptable, providing, of course, that the underlying training ('observed') data are accurate.

The cattle prediction mirrors the observed distribution well (Figure 1 top), and picks out both major foci (e.g. East African and Zimbabwe highlands, Tanzania, semi-arid and dry sub-humid West Africa) and smaller concentrations such as in the Gezira, the Mali Delta, and south eastern Zambia. Relatively high resolution spatial data that exist in the observed survey data for Nigeria and Botswana tend to be smoothed out by the regression methods used to generate the predicted map. Some of the contrasts between observed and predicted maps are due to minor differences between observed and predicted values falling into different mapping classes. There are also some minor anomalies in northern Chad, where very high predicted densities are obviously false, and are caused by extreme predictor values. The major predictor is human population density, which is primary in 30% of the equations.

PAATIS provides two zoning layers - ecozones (used to subdivide the predictive analyses described above) and farming systems. Both are intended to show areas with similar (eco-climatic or agricultural) characteristics. They were produced using a statistical clustering technique available within the ADDAPIX software produced by FAO. The ecozones were defined using elevation, and remotely sensed imagery relating to temperature, rainfall, vegetation cover and vapour pressure deficit. The farming systems were identified using cattle, cropping and human population levels as well as elevation.

The layers for Forests and for National Parks and Reserves were obtained from FAO and NRI. The latter boundaries are for IUCN categories I to IV and supplemented by ERGO archive data for Botswana and Nigeria. It is stressed that these data are incomplete – for example neither the gazetted areas of South Africa, nor the forest information for the SADC countries are included.

Major rivers and river basins were obtained from USGS EROS (*edcdaac.usgs.gov/gtopo30/hydro/africa.html*), and roads and administrative boundaries FAO, and NRI.

Digital Elevation Model (DEM) data were obtained from the GTOPO30 1km resolution elevation surface for Africa, produced by the Global Land Information System (GLIS) of the United States Geological Survey, Earth Resources Observation Systems (USGS, EROS) data centre.

The length of growing period (LGP) and mammal bio-diversity layers are derived from low resolution data by the same predictive procedures used for cattle and cropping levels. The training data for LGP were provided by FAO AGL, whilst those for mammal distributions were extracted from the African Mammal Database compiled by the Istituto di Ecologia Applicata in Rome.

Rainfall is derived from the public domain maps published by Cramer and Leemans, provided by the Environmental Change Unit of Oxford University.

PAATIS also provides a background layer the Normalised Difference Vegetation Index (NDVI), a widely used measure of 'greenness' allied to vegetation cover. This layer is derived from Advanced Very High Resolution Radiometer (AVHRR) satellite imagery, from the Pathfinder Program, initially supplied by the NASA Global Inventory Monitoring and Modelling Systems (GIMMS) group, and further processed by the TALA research group.

DESIGN AND FUNCTIONALITY

The GIS was designed using ArcView® (v 3.1) with the Spatial Analysis® Extension (v1.1) for analysis and data manipulation of raster images. This software was chosen because of its object-oriented scripting language (Avenue®) that allows complete customisation of the Graphical User Interface (GUI). The PAATIS design is modular so as to facilitate easy updating of existing layers, addition of new layers and customisation of the user interface to incorporate new modelling and statistics options. This provides a variety of options for future development.

The GIS was organised into three layers which correspond to three stages of the decision pathway: i) defining and focusing on the study area, ii) mapping priority control areas defined by cattle and crop levels and presence or absence of tsetse and iii) mapping and evaluating predicted impact of control. At each level, the user is provided with various options: mapping tasks to customise the view; analysis tasks to extract geographical information and statistics; navigation tasks to personalise the overall decision pathway;and export tasks to produce graphical or tabular outputs for use in external software.

Statistics, priority control maps and impact maps can be generated or calculated for a geographic area or for a shape defined by the user. Local statistics can be calculated for all data layers marked * in Table 1 and cross tabulated statistics by Country and/or Ecozone and/or Farming System can be calculated for all those marked [†]. All tabular results can be exported to table format files or straight to an Excel spreadsheet.

KNOWLEDGE BASE AND RESOURCE INVENTORY

The Knowledge Base consists of over 6000 records, with abstracts where applicable, extracted from the last 8 years of the Tsetse and Trypanosomiasis Information Quarterly (TTIQ). The Resource Inventory has been compiled from miscellaneous sources, with notable assistance from the annual reporting of the FAO Liaison Officers Network. Table 2 outlines the categories available to the user at the national level.

The Knowledge Base and Resource Inventory components are stored in Microsoft Access 97 files and are accessed within the customised user interface of the GIS through query windows designed in Microsoft VisualBasic 6. The process of retrieving information from the Knowledge Base and Resource Inventory can either be completely independent of the GIS or linked to the selection of a country. Queries of the Knowledge Base can be made by Author, Country, TTIQ Categories, Keywords, and publication year. Queries of the Resource Inventory are country-based.

Table 2: Content of the Resource Inventory at the national level.

Category	Sub-categories
Contacts	Contact details of FAO Liaison Officer
	List of Institutes involved in tsetse and trypanosomiasis
	Details of staff resources
Overview	A brief overall report on the tsetse-trypanosomiasis
	situation during reporting period
Cattle	Total cattle
	Overview map of cattle distribution
	Number of cattle at risk
	Total average meat production ('000 mt)
	Total average milk production ('000 mt)
	Estimate of cattle not kept because of tsetse
Cropping	Total land area (sq km)
	Total arable land (sq km)
	Total permanent crops (sq km)
	Overview map of land in cultivation
Tsetse	Total area of tsetse (sq km)
	Proportion of tsetse area (as % of total land area)
	Overview maps of fly distribution & control operations
	Details of current tsetse control programmes
Animal trypanosomiasis	Description of present situation of trypanocidal drug use and pour on applications
Human Sleeping Sickness	Brief description of present HSS situation
Financial	Annual financial input to all projects
	Proportion of funding from internal/external sources
	List of funding Institute(s)/Organisation(s)
	List of receiving Institute(s)/Organisation(s)
FAO Liaison Officer Report	Annual report with tables and figures, if applicable
Broader Issues	Historical perspective, Terrain, Climate, Natural hazards, Environment and Land use
Economy Overview	Overview of recent economic situation

PAATIS OUTPUTS

Simultaneous display of the 3 primary PAATIS components presents the user with a novel and powerful approach of combining spatial GIS information, and complimentary text based information. Both the Knowledge Base and Resource Inventory are geo-referenced by country allowing for cross referencing relevant material.

Two methods of calculating tsetse impact are provided in the GIS Level 3. Users have been provided with a means of setting their own impact levels (both positive and negative) which the programme then applies to the cattle and cultivation data. Secondly, a predicted impact of the removal of the fly on cattle (Figure 3) and cropping levels, is provided. This is derived using the predictive equations for cattle and crops, within which the number of tsetse species is set to zero which gives an estimate of the levels of these parameters in the absence of tsetse. This is subtracted from the original prediction to give a predicted impact.

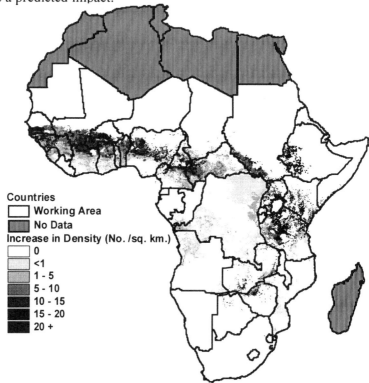

Figure 3: Predicted impact of tsetse removal on cattle densities (PAATIS, 2000)

It is fully appreciated that this is a rather simplistic way of calculating impact, as it assumes that all tsetse species have a similar effect on agriculture, and that this is constant for all areas and agro-ecological zones. The method does not take into account any possible increase in human population (and its possible impact on agricultural levels), nor does it predict any new agriculture where there is currently none. As such the predicted impact is likely to be an underestimate. The method does, however, highlight the areas where some agriculture is already found within the tsetse belt - and thus likely to be most immediately affected by tsetse control operations. In West Africa such areas are exemplified by parts of Burkina Faso and Mali; and in the East, Ethiopia, Tanzania and the Lake Victoria Basin.

Whilst an estimation of the possible change in cattle and cultivation due to the removal of tsetse is central to the definition of impact, it is the economic value of these changes that often determines the feasibility of control operations. It was impractical to assign fixed 'dollar values' to the impact predictions provided so users have been provided with the facility for assigning their own (currency independent) values to a hectare of cultivation and a cattle density of one animal per square kilometre, which are then applied to the impact predictions, thereby providing an indication of economic impact that may be adjusted to local conditions.

Results from these impact maps can be extracted by any geographical region or shape, but for the purposes of comparison, Figure 4 provides a country inventory of predicted change in cattle numbers assuming the removal of all tsetse species. For the 37 countries included in the analysis, present data indicates that there are approximately 175 million head of cattle, of which 45 million are within the area of tsetse infestation. From the predicted cattle impact map the PAATIS calculates that there would be an increase of approximately 50 million head of cattle (effectively a 100% increase in actual numbers) if all tsetse were removed.

Figure 4: Predicted increase in cattle numbers if tsetse removed. (PAATIS, 2000)

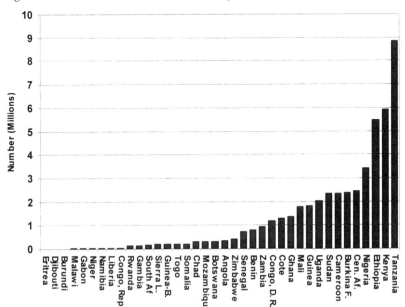

DISCUSSION AND POTENTIAL IMPROVEMENTS
Future improvement in decision support for tsetse control strategies
The cattle and cultivation impact predictions represent the agro-economic dimension of one approach towards rational decision support for tsetse control strategies. Improved methodology, better quality training data and observed changes in cattle/cropping after control campaigns would lead to improved impact modeling, as would the incorporation of socio-economic information. However, it is clear that at least two additional factors require consideration - agro-ecological setting and an index of the technical feasibility of tsetse control. The agro-ecological setting reflects the suitability for mixed farming development which can partially be defined by the length of growing period. Refinement of this layer could be through a land pressure index defined using the number of cattle or cultivated hectares per 100 people. Preliminary attempts at this type of analysis are given in FAO (1999). Definition of a technical feasibility score would be more difficult. Analysis of the degree of isolation of tsetse pockets and spatial patterns could provide an insight into fragmentation of tsetse populations, but this would be dependent on better tsetse species distribution and abundance maps, such as those presented at *ergodd.zoo.ox.ac.uk/tseweb/index.htm*. This in turn could lead to the creation of cattle/cropping impact maps by individual tsetse species or groups, thereby addressing the problem of the current approach that assumes all tsetse species have a similar impact on agriculture. In effect, this would lead to the creation of a scoring index for each tsetse species which could be generated for individual ecozones, farming systems and/or land pressure indices.

Future of the PAATIS software package.
Continental scale information could simply be replaced by a country or project specific geographical area with associated higher resolution data layers. Existing cattle layers could be supplemented with data for other animal species or trans-boundary disease occurrence, such as addressed by the EU funded PACE and FAO EMPRES. The economic importance of, for example, production systems, milking and calving offtake, could be incorporated into the analysis and impact modelling. A prototype of this approach has recently been tested for Kenya by FAO AGAH.

Whilst it is acknowledged that there are limitations in the validity of existing continental tsetse maps, it must be noted that there are currently developments in the field of predicting tsetse distribution and abundance using satellite imagery, which in turn would lead to more detailed and accurate analysis of impact by individual tsetse species or groups. A distinct advantage of the spatial approach within this customised environment, is the ability to obtain statistics, not only from administrative boundaries, but from a user defined shape reflecting a more realistic trans-boundary approach. For example, the extraction of any of the statistical outputs provided in the PAATIS menu could be applied to a natural watershed or some other terrain related system. In addition, statistics can also be calculated by selecting a cross-tabulation output by LGP category, Ecozone or Farming System. This not only allows more flexible data queries than available in the majority of information systems, but provides a decision tool that can be targeted precisely to specific project requirements. There is thus substantial potential for expanding the PAATIS to incorporate a range of additional data, utilising its modular construction, to become more widely applicable as a source of agro-ecological and epidemiological information. Finally, PAATIS is intended to provide an extensive and standardised set of agro-ecological and tsetse related data to a wide audience. It is hoped that this will stimulate recipients of the system to assist PAAT in updating and revising the many data layers included, as well as to refine and improve the capabilities of the system itself.

ACKNOWLEDGEMENTS
We would like to acknowledge the support and enthusiasm of all members of the PAAT Secretariat, Committee and Advisory Group Coordinators. Invaluable contributions have also been received from the FAO Liaison Officers Meetings coordinated by George Chizyuka (FAO Accra). Finally PAAT would not exist without the commitment of Brian Hursey, to whom we are all grateful. The PAATIS was funded by FAO and the Department of International Development, UK (DFID).

REFERENCES
FAO (1996a): Livestock Geography: A demonstration of GIS techniques applied to Global Livestock Systems and Populations. Consultancy Report by ERGO Ltd to the Animal Health Division of the Food and Agriculture Organisation of the United Nations, Rome.

FAO (1996b): Livestock Geography II: A further demonstration of GIS techniques applied to Global Livestock Systems, Populations and Productivity. Consultancy Report by ERGO Ltd to the Animal Health Division of the Food and Agriculture Organisation of the United Nations, Rome.

FAO (1996c): Towards Identifying Priority Areas for Tsetse Control in East Africa. Consultancy Report By D. Rogers and W. Wint, prepared by TALA and ERGO Ltd for the Animal Health Division of the Food and Agriculture Organisation of the United Nations, Rome.

FAO (1997): Ecozones, Farming Systems and Priority Areas for Tsetse Control in East, West and Southern Africa. Consultancy Report by W.Wint, D. Rogers and T.

Robinson, prepared by ERGO Ltd and TALA for the Animal Health Division of the Food and Agriculture Organisation of the United Nations, Rome.

FAO (1998): Prediction of Cattle Density, Cultivation Levels and Farming Systems in Kenya. Consultancy Report by W.Wint and D.Rogers, prepared by ERGO Ltd and TALA, for the Animal Health Division of the Food and Agriculture Organisation of the United Nations, Rome.

FAO (1999): Agro-Ecological Zones, Farming Systems and Land Pressure in Africa and Asia. Consultancy Report by W. Wint, J. Slingenbergh and D. Rogers, prepared by ERGO Ltd And TALA for the Animal Health Division of the Food and Agriculture Organisation of the United Nations, Rome.

FAO (2000) Predicted Distributions Of Tsetse In Africa Report and database by W.Wint and D. Rogers prepared by Environmental Research Group Oxford Ltd and TALA Research Group, Department of Zoology, University of Oxford, for the Animal Health Service of the Animal Production and Health Division of the Food and Agriculture Organisation of the United Nations, Rome, Italy.

Kerby, P. (1997): Development of the FAO/PAAT Information System. Consultancy Report by Peter Kerby to the Animal Health Division of the Food and Agriculture Organisation of the United Nations, Rome.

Gilbert, M. (1999): Illustrated Guide on Linking Animal Productivity Information with GIS-based Livestock Population Inventories and the Preparation of Cattle Production Atlas for Kenya. Consultancy Report by Marius Gilbert to the Animal Health Division of the Food and Agriculture Organisation of the United Nations, Rome.

EFFECTS OF CLIMATE, HUMAN POPULATION AND SOCIO-ECONOMIC CHANGES ON TSETSE-TRANSMITTED TRYPANOSOMIASIS TO 2050

J.J. McDermott, P.M. Kristjanson, R.L. Kruska, R.S. Reid, T.P. Robinson, P.G. Coleman, P.G. Jones and P.K. Thornton
International Livestock Research Institute, P.O. Box 30709, Nairobi, Kenya, International Centre for Tropical Agriculture, AA6713, Cali, Colombia (Jones)

ABSTRACT This chapter explores the impacts of climate change, human population growth and expected disease control activities on tsetse distribution and trypanosomiasis risk in five agro-ecological environments in sub-Saharan Africa to 2050. These changes will tend to contract areas under trypanosomiasis risk continent-wide; however, this trend will not be uniform. The greatest decrease in the impacts of animal trypanosomiasis will occur in the semi-arid and sub-humid zones of West Africa, where the climate will be drier, human population will increase and disease control will have greater impacts. The risk of animal trypanosomiasis will also decline in many but not all areas of Ethiopia and eastern and southern Africa. The disease situation in the humid zone of central and western Africa will be less changed. Sleeping sickness, particularly the gambiense type, will continue, as now, to be a major problem, if concerted control efforts are not implemented.

Key words trypanosomiasis, tsetse, climate change, human population growth

To predict the future, historians argue, we need to learn from the past. If so, those of us bold enough to make predictions on future tsetse and trypanosomiasis trends are in trouble. The patterns of human and animal tsetse-transmitted trypanosomiasis in sub-Saharan Africa (SSA) over the 20th century present a murky picture on which to base predictions for the 21st century. Animal trypanosomiasis, commonly called *nagana*, remains an important problem in SSA, where control efforts and their impacts have been uneven. Large-scale publicly funded vector control efforts have cleared no more than an estimated 2% of tsetse-infested land (Budd, 1999). Farmer-based private control efforts – overwhelmingly based on chemotherapy and in recent years also on pour-on insecticides - have increased. However, their large-scale impacts have not yet been seen. Human trypanosomiasis, known commonly as sleeping sickness, has risen in recent years to levels not seen since the 1930s and is now a serious economic development as well as public health concern. Hardest hit are those areas with civil unrest, where basic community health efforts and other social infrastructure have broken down. We can thus exploit no pattern of success in reduction of either animal or human trypanosomiasis for estimating future impacts of control efforts.

Nevertheless, our knowledge of this relatively short 100-year history provides two important clues as to what lies ahead. Past trends indicate that socio-economic factors, particularly development, human population growth and distribution, and evolving agricultural and livestock production systems, will increasingly determine the incidence and distribution of sleeping sickness and nagana and that biophysical factors, both influenced by and independent of these socio-economic factors, will remain important determinants.

In this chapter, we focus on predicting changes in tsetse and trypanosome distribution and the impacts of those changes in SSA. Two global drivers, climate and human population, will be useful predictors of these changes. Climate and human population directly influence trypanosomiasis risk and may be correlated with other useful predictors such as habitat (vegetation, especially tree and bush cover) and livestock numbers. We bear in mind, however, that unforeseen events can dramatically change disease trends. The devastating cattle rinderpest epidemic that swept across the continent in the late 19th century, for example, radically changed the patterns and impacts of animal trypanosomiasis. One hundred years later, widespread and protracted civil wars led to a resurgence in sleeping sickness occurrence while the AIDS epidemic greatly reduced resources available for sleeping sickness control.

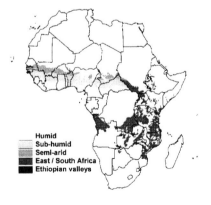

Figure 1. Broad agro-ecological areas of tsetse and trypanosomiasis risk.

Our predictions of the distributions and impacts of tsetse flies and trypanosomiasis over the next half-century focus on changes within agro-ecological environments. Because we are emphasizing broad trends, we consider five such environments, (1) the humid (2) sub-humid (3) and the semi-arid zones of West and Central Africa; (4) the Ethiopian river valley lowlands; and (5) the tsetse-infested meat-dairy livestock areas in eastern and southern African. These zones (Figure 1) are described in more detail elsewhere (McDermott and Coleman, 2001). We selected these zones using five criteria: presence of tsetse, geographic region, agro-ecological environment (rainfall, temperature and elevation), trypanosomiasis risk, and differences in current major control strategies. Within these zones we analyzed changes in tsetse presence and trypanosomiasis risk, due both to continental climate and human population drivers and to system-specific changes.

Effects of climate change on tsetse distribution and impact

Over the millennia, climate change has had a profound influence on African populations, as elsewhere, affecting both their size and distribution and their natural resource management and agricultural strategies (Reader, 1997). Typically, people have responded more to climate change in "marginal" areas such as the Sahel. To predict future changes in tsetse distribution and impact, we use the length-of-growing period (LGP) as an indicator of the impact of climate change on natural resource attributes (particularly tree cover), agricultural activity and tsetse distribution and trypanosomiasis risk. The LGP defines a period when crop production is possible based on temperature and moisture limitations (roughly when precipitation exceeds half the potential evapotranspiration); in most of SSA agriculture is determined by moisture (FAO, 1978).

A current LGP surface for SSA was estimated based on long-term monthly climate normals (rainfall, daily temperature, and daily temperature diurnal range) from over 7000 stations (Jones, 1987). A 10 minutes-of-arc surface was fitted, based on the NOAA data set TGOP006 (NOAA, 1984) using inverse-square distance weights for spatial interpolation and correction for elevation effects. For each grid cell, a simple daily water balance model was run, interpolating the monthly climate normals to daily values using a fast Fourier transform (Jones, 1987). To derive a predicted LGP surface for 2050, mean values of maximum and minimum temperature and precipitation for 2041-2070 were obtained from the

Intergovernmental Panel on Climate Change (IPCC) Data
Distribution Centre on the worldwide web (http://ipcc-
ddc.cru.uea.ac.uk/). These predictions are based on a greenhouse-
gas-only experiment conducted at the Hadley Centre using the
Unified Model (Cullen, 1993) at a resolution of 2.5 by 3.75 degrees of
latitude and longitude. These data were then interpolated to the 10-
minute pixel size (Jones and Thornton, 2000) and the water balance
model rerun to produce the 2050 LGP surface. Figure 2 shows the
current 2000 LGP surface and the 2050 and 2000 difference map.
Major reductions in LGP are predicted for areas of West Africa,
southern Sudan, Uganda and some lowland areas of Ethiopia and
increases in south-eastern Kenya, north-eastern Tanzania, southern
Cameroon and some highland areas of Ethiopia.

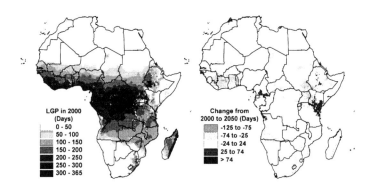

Figure 2. Distribution of length-of-growing period (LGP) in 2000
(left) and predicted changes in LGP to 2050 (right).

The predicted changes in LGP were then used to predict changes
in the distribution of the three main groups of tsetse flies (*Glossina
spp.*) -- *Morsitans* (savanna), *Fusca* (riverine) and *Palpalis* (forest).
The first step was to estimate the probability of different LGP values
supporting tsetse flies by comparing the LGP surface to the current
presence or absence of each tsetse group in 10-minute spatial grids.
Then, by Bayes' theorem, the probability of tsetse presence or
absence across the range of LGP values was calculated and graphed.

For a single variable such as LGP, thresholds are simply the
values at which the probability density functions for presence and
absence cross. A better summary approach, the method of optimal
threshold distribution functions (OTDF) (Robinson *et al.*, 1997), is to
compare the cumulative distribution functions for LGP given that

flies are present or absent, respectively. The OTDF gives the
probability that flies are present above a given value and absent below
it. Figure 3 shows OTDF plots across a range of LGP values for each
group of tsetse, assuming equal *a priori* probabilities. In

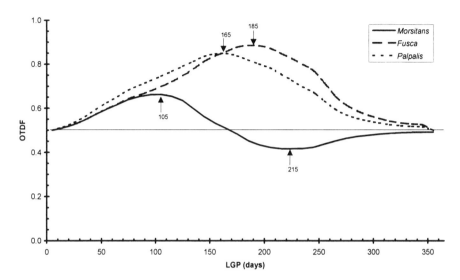

Figure 3. Optimal distribution functions for Morsitans, Fusca *and* Palpalis *tsetse groups.*

general, flies should be present when the OTDF plots are falling and
absent when they are rising. If the OTDF reaches either one or zero
the prediction is perfect; if it is flat at 0.5 the variable has zero
predictive power.

Two things are immediately obvious from Figure 3. First, the
power of LGP alone to predict tsetse distribution is much greater for
the *Fusca* and *Palpalis* groups than for the *Morsitans* group. Second,
while there is a single lower threshold of LGP for the *Fusca* and
Palpalis groups of flies (LGP = 185 and 165 respectively) above
which habitat should be suitable, for the *Morsitans* group there is a
lower (LGP = 105) and upper (LGP = 215) threshold between which
habitat should be suitable.

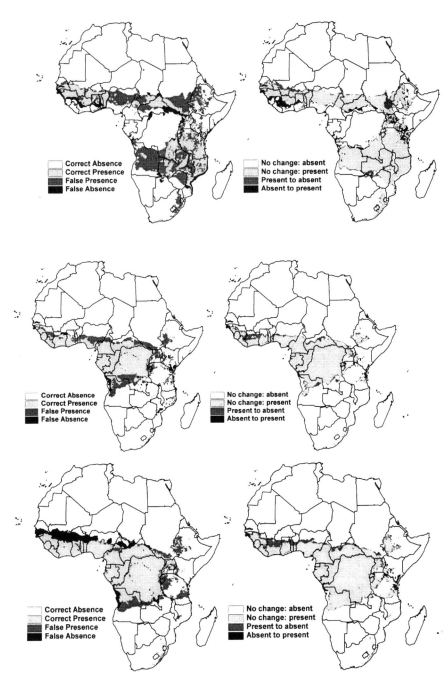

Figure 4. Model predictions compared to current distribution of Morsitans *(top left),* Fusca *(center left) and* Palpalis *(bottom left) tsetse groups and predicted changes in distribution to 2050 (* Morsitans *(top right),* Fusca *(center right) and* Palpalis *(bottom right)).*

Having determined the OTDF of LGP for each group we can then use these values to predict the spatial distribution of tsetse based on LGP now and in 2050 and thus make inferences on the likely change in tsetse distribution based on changes in LGP. Figure 4 shows, for each group, the actual current distribution (modified from Ford and Katondo, 1977) and the predicted changes in distribution to 2050 showing areas of continued presence or absence, areas that have become suitable, and areas no longer suitable. For the *Morsitans* group, predictions based solely on LGP are poor. Nevertheless, general predicted trends are a decrease in habitat area suitable for tsetse along the northern front of the west African fly belt, over a large area in southern Sudan and in southern Zambia and an increase in habitat suitability along the southern front of the west African fly belt and in scattered parts of Kenya, Tanzania, Uganda and Ethiopia. For the *Fusca* group, LGP is a much better predictor. In general, a decrease in habitat suitability along both the northern and southern fronts of the distribution and an increase in habitat suitability in scattered parts of East Africa are predicted. Likewise, for the *Palpalis* group, the prediction based on LGP is very good, with a similar pattern of change expected.

Effects of human population growth on tsetse distribution

While changes in climate are expected to cause gradual changes in tsetse distribution over the next 50 years, changes due to increases in human population will probably be more rapid. The crude associations between human population growth and decreasing tsetse distribution, particularly in Nigeria (Jordan, 1986), have led to speculation that high population growth could eliminate tsetse flies from large areas of Africa, irrespective of control activities. The assumption, still requiring confirmation, is that human population density is highly correlated with the area of tsetse habitat cleared for cultivation. This prospect encourages those working to eliminate trypanosomiasis and frightens those who view the tsetse fly as the guardian of African ecosystems (e.g. Ormerod, 1990). A recent paper by some of us (Reid *et al.*, 2000) explored the spatial associations between human population and tsetse distributions to 2040. These predictions have been updated to 2050 for this chapter.

Moderate-resolution human population GIS data layers have been developed (Deichmann, 1996) based on population density data from 1960, 1970, 1980 and 1990. Predictions for 2000 were made by extrapolation from 1990 estimates using 1980-90 cell-by-cell growth rates (Reid *et al.*, 2000), because these, rather than earlier rates, were considered to best reflect likely demographic trends (minimal rural-rural migration and continued high rural-urban migration (Foote *et al.*, 1993)). In extrapolating human population from 2000 to 2050, cell-by-cell growth rate estimates from 1980-90 were adjusted downward by country, proportional to "medium" UN country-level projections for 2050 to account for expected lower fertility and higher mortality due to AIDS. The 2050 predicted spatial patterns of human population thus developed reflect major expected migrational and demographic trends and match UN country-level population projections.

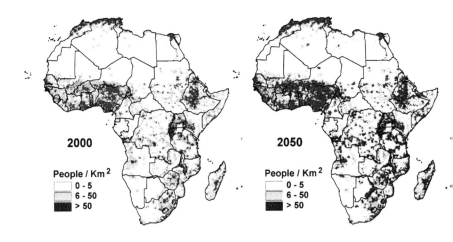

Figure 5. Distribution of human population in 2000 and 2050.

To predict the impacts of human population growth on the distribution of tsetse, the following considerations and assumptions were made (these are described and developed in more detail in Reid *et al.*, 2000). The first was that the responses of the different tsetse groups to human population pressure would differ. The impacts of human population on *Fusca* (riverine) flies were considered minimal, because these flies adapt well to many peri-domestic habitats (e.g. Jordan, 1986). Thus, future human population – tsetse scenarios were developed only for the *Palpalis* and *Morsitans* groups. The second, simplifying, assumption was to classify the impacts of human population on tsetse distribution into 3 categories: (1) low human population density having no effect on tsetse populations; (2) moderate human population density causing a decline in tsetse populations and (3) high human population density eliminating Palpalis and Morsitans populations. Third, it was considered that thresholds for these categories might vary from place to place. Two scenarios, liberal and conservative, were developed. For the liberal scenario (maximal human impact on tsetse), the human population thresholds for no effect and extinction were 15 and 39 people km^{-2} and for the conservative (minimal impact) scenario, 30 and 77 people km^{-2}, respectively.

We predict human population growth will have important impacts on tsetse distributions between 2000 and 2050 (see Figure 6). Our results tell us, however, that tsetse will remain in high numbers over more than 50% of their current distribution and in moderate numbers over an additional 20%. The greatest decline in overall tsetse populations will occur in West Africa, with smaller areas of decline in eastern, central and southern Africa (Lake Victoria Basin, north-western Uganda, coastal areas of eastern and southern Africa and large areas of central Africa). Forest (*Palpalis*) flies are likely to disappear in parts of eastern and southern Africa, coastal West Africa, and several large patches of central Africa. While the rate of decline will be faster in the liberal than conservative scenario, the same locations will be affected. Despite the assumption of little human impact on riverine (*Fusca*) tsetse distribution, there will no doubt be some decline influenced by a rising human population, particularly in drier areas where isolated riparian forests come under threat from human use – a likely scenario in West Africa.

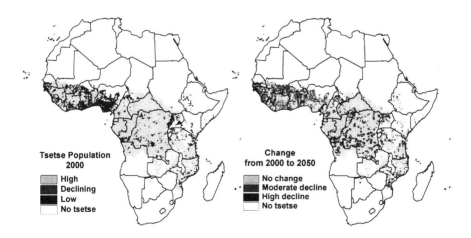

Figure 6. Current tsetse distribution (left) and conservative scenario for the change in tsetse population due to human population growth to 2050 (right).

Human population growth alone is likely to contract the area infested by tsetse by some 10%, compared with approximately 2% cleared by tsetse control operations since 1970 (Budd, 1999). While tsetse-infested areas in 2050 will have the same number of humans and livestock as those currently at risk, this will represent a much smaller proportion of the total African human and livestock populations.

Variations in future trypanosomiasis risk and control in major environmental / agricultural zones

Table 1: Variations in trypanosomiasis risk and impact in SSA.

Zone	Main systems	Principal losses	Products affected	Market access	Control Methods
Humid West and Central Africa	AP, LS, MOS	Mortality, reproduction, breed options, milk, meat, sleeping sickness	milk, meat	L H MOS	Trypano-tolerance
Sub-humid West Africa	AP, P	Mortality, reproduction, breed options, milk, meat, draught, control costs	milk, meat, cotton, maize, groundnuts	M	Trypanocides Trypano-tolerance
Semi-arid West Africa	P, AP	Restrictions to movement / grazing; milk, meat, control costs	milk, meat, millet, sorghum	M	Trypanocides
Valley lowlands Ethiopia	AP, P	Draught, mortality, reproduction, breed options, milk, meat, control costs	meat, teff, wheat, milk	L	Trypanocides
Eastern and Southern Africa	MOSS LS AP, P	Milk, meat, mortality, reproduction, breed options, control costs; draught, sleeping sickness	milk, meat, maize	M-H	Trypanocides Vector Control

Main systems: P=pastoral; AP=agro-pastoral (mixed crop-livestock, subsistence-oriented); LS=large-scale commercial; MOS=Market-oriented smallholder.
Market access: L=low; M=medium; H=high.

To assess future trends in the impacts of animal and human trypanosomiasis impact (see also Kristjanson *et al.*, 1999), we combined the continental-level influences of climate and human population with system-specific characteristics. Important characteristics of the five broad zones shown in Figure 1 are listed in Table 1.

We expect the most dramatic changes in trypanosomiasis risk and impact to occur in the sub-humid and semi-arid zones of West Africa. Climate and to a lesser extent human population growth will have major effects in the semi-arid zone, reducing transmission to small riverine pockets. In these pockets, both tsetse (almost exclusively *Glossina palpalis gambiensis*) and trypanosomes (*Trypanosoma vivax* and *T. congolense*) can be controlled, but they will be difficult to eradicate with conventional techniques if sensitive riparian habitats can be preserved. In the sub-humid zone, changes in both human population and climate will greatly reduce risk. As market-oriented systems for cotton and other cash crops develop, we foresee farmers improving their control of trypanosomiasis using strategic vector control and chemotherapy. Publicly funded area-wide vector control is likely to have greatest economic impact in this zone and could result in eliminating tsetse in isolated areas where tsetse re-invasion can be avoided.

Changes in the humid zone of West and Central Africa are likely to be more gradual and less dramatic. Climate change and human population growth will decrease trypanosomiasis risk in patchy areas only. Thus, gambiensis sleeping sickness and animal trypanosomiasis risk will be similar for most areas and age-old strategies for reducing the risks of living with tsetse and trypanosomes will continue. For animal trypanosomiasis, we expect that current low numbers of cattle and small ruminants will increase, particularly in the coastal strip of West Africa. Farmers will respond to increasing consumer demand for meat and milk in two ways -- with improved selection and management of trypanotolerant breeds and by intensifying production in peri-urban areas with lower disease risk. The extent of improved livestock production on a broader scale will largely depend on the resolution of the numerous civil conflicts currently plaguing the region. These conflicts will also inhibit major efforts to control gambiensis sleeping sickness. As there is little prospect for a significant decrease in natural disease risk, concerted international action provides the only realistic hope of alleviating the current sleeping sickness crisis.

In Ethiopia, tsetse and trypanosomiasis risk will remain important constraints to livestock and crop production. A net reduction in tsetse risk in current risk areas with perhaps a slightly broader distribution would be predicted based on climate change (enhanced suitability for tsetse at higher elevations) and human population growth (declining tsetse numbers in increasingly populated valleys). Thus, successful control efforts will be required for major reductions in risk. Two general approaches are envisaged: publicly supported, large-scale vector control and farmer-based control that integrates the use of insecticides and strategic treatment with trypanocides, which will be increasingly less efficacious.

The eastern and southern African situation has been and will be the most heterogeneous. In areas of highest population growth and highest market access, both population pressure and conventional control by farmers will reduce trypanosomiasis impacts on livestock and crop production. Trypanosomiasis will continue to be a major constraint to production in lowland areas, particularly along the coast, where the effects of human population growth in some areas will be balanced by a more favourable climate for tsetse in others. Some focal areas of rhodesiense sleeping sickness will remain, but these could be largely controlled by 2050 through application of conventional strategies.

Less Predictable Trends

In closing, we wish to revisit the crucial role of socio-economic factors in shaping the future pattern of tsetse and trypanosomiasis risk and impact in SSA. As described above, biophysical factors, such as climate changes and human population growth will be important, but these are not likely by themselves or together to cause dramatic continental changes in trypanosomiasis patterns. Two broad socio-economic factors are likely to be much more influential – the ability to resolve conflicts and the evolution of African economies.

Conflicts have been and are likely to remain the key determinant of the current sleeping sickness epidemic in Central Africa. Likewise, there is huge potential for conflicts between pastoral and agro-pastoral peoples in both East and West Africa. In the absence of sound and potent local, national and regional policies and actions, such conflicts will seriously compromise coordinated tsetse and trypanosomiasis control programs.

The growth rates of African economies and the evolution of markets, institutions and policies will constitute the main influences on future impacts of trypanosomiasis on livestock production. Trypanosomiasis will probably be controlled most effectively and sustainably in those areas of East and West Africa with better infrastructure and market access, provided strong incentives exist for individuals to invest in control programs. Such incentives are likely to be most successful if linked to efficient input-output markets – for example, for milk and meat marketing to urban areas and for cotton and other cash crops in systems in which animal traction is an essential input. Farmer organizations could help by facilitating output sales and animal health and production inputs. While such private initiatives will increase in areas where economies are growing and infrastructure and policy support strong, the worry is that public services will not be supplied to less able and less advantaged communities. The current continental decline in provision of public services such as community health and agricultural extension will likely continue if there is no significant change in current trends of poor public-sector management and declining donor funding. A key challenge in the medium to long term will be to identify the most appropriate mix of private and public investment in trypanosomiasis control and the most effective means of instituting effective public-sector programs when they are warranted. Unless reversed, the current trend of declining public-sector investment is likely to undermine efforts to deliver and sustain large-scale tsetse and trypanosomiasis control programs in the foreseeable future.

In summary, without new, concerted, and innovatively delivered control efforts, the impacts of animal trypanosomiasis will gradually decline in some endemic areas but continue to effect almost the same number of livestock as today. Likewise, without major changes in the current social, political and economic situation, we fear that conflicts and poor infrastructure will continue to thwart efforts to treat and control sleeping sickness. Thus, trypanosomiasis, diminished in some areas, is likely to remain an important problem for Africa to 2050 and beyond.

References

Budd L.T. 1999. DFID-funded tsetse and trypanosomiasis research and development since 1980. Vol. 2 – Economic Analysis. DFID, London, 123p.

Cullen M.J.P. 1993. The unified forecast/climate model. *Meteorological Magazine* **122**: 81- 95.

Deichmann U. 1996. Africa Population Database, Digital Database and Documentation. University of California, Santa Barbara, USA.

FAO (1978). Report on the Agro-Ecological Zones Project. Volume 1, Methodology and Results for Africa. World Soil Resources Report 48, Food and Agriculture Organisation of the United Nations, Rome, Italy.

Foote K.A., K.H. Hill, and L.G. Martin (eds.). 1993. Demographic change in sub-Saharan Africa. National Academy Press, Washington D.C.

Ford J., and K.M. Katondo. 1977. Maps of tsetse fly (*Glossina*) distribution in Africa, 1973, according to sub-generic groups on a scale of 1:5,000,000. *Bulletin of Animal Health and Production in Africa* **15**: 188-193.

Jones P.G. 1987. Current availability and deficiencies in data relevant to agro-ecological studies in the geographic area covered by the IARCS. *In* Agricultural Environments: Characterisation, classification and mapping, A.H. Bunting (ed.). CAB International, Wallingford, UK. p. 69-83.

Jones P.G., and Thornton P.K. 2000. MarkSim: Software to generate daily weather data for Latin America and Africa. *Agronomy Journal* **92**: 445-453.

Jordan A.M. 1986. Trypanosomiasis Contr
ol and African Rural Development. Longman, London, 357p.

Kristjanson, P.M., B.M. Swallow, G.J. Rowlands, R.L. Kruska, and P.N. de Leeuw. 1999. Measuring the costs of African animal trypanosomiasis, the potential benefits of control and returns to research. *Agricultural Systems* **59**: 79-98.

McDermott J., and P. Coleman. 2001. Comparing apples and oranges – model based assessment of different tsetse-transmitted trypanosomiasis control strategies. *International Journal of Parasitology* (in press).

NOAA (National Oceanographic and Atmospheric Administration). 1984. TGP-OO6 D. Computer compatible tape. NOAA, Boulder, Colorado, USA.

Ormerod W.E. 1990. Africa with and without tsetse. Insect Science Applications **11**: 455-461.

Reader J. 1997. Africa: A Biography of the Continent. Hamish Hamilton, London, 803p.

Reid R.S., R.L. Kruska, U. Deichmann, P.K. Thornton, and S.G.A. Leak. 2000. Human population growth and the extinction of the tsetse fly. *Agricultural Ecosystems & Environment* **77**: 227-236.

Robinson T.P., D.J. Rogers, and B. Williams. 1997. Univariate analysis of tsetse habitat in the common fly belt of Southern Africa using climate and remotely sensed vegetation data. *Medical and Veterinary Entomology* **11**: 223-234.

TSETSE VECTOR BASED STRATEGIES FOR CONTROL OF AFRICAN TRYPANOSOMIASIS

S. Aksoy

Department of Epidemiology and Public Health, Section of Vector Biology, Yale University School of Medicine, 60 College St., 606 LEPH, New Haven, CT 06510, USA.

ABSTRACT

The application of recombinant DNA technologies for molecular genetic approaches promises to bring about new strategies for control of vector-borne-diseases. These approaches can also enhance the existing tools. Here, one application of this technology is presented where parasite refractory insects are engineered that can then be spread to replace their susceptible counterparts in the field to reduce disease transmission. The approach presented here utilizes the symbiotic bacteria that are naturally harbored in tsetse to express foreign genes. As these naturally harbored organisms reside in the same tissues as trypanosomes, the expression of anti-trypanosomal products in these bacterial symbionts can adversely effect parasite biology. The use of *Wolbachia* symbionts, which are known to induce phenomena such as cytoplasmic incompatibility, is discussed as a potential gene driving system.

Key words: *Glossina*, trypanosomiasis, symbiosis, transgenesis, vector control, *Sodalis, Wigglesworthia, Wolbachia*

INTRODUCTION

Despite decades of research with vector-borne parasites and their development in mammalian hosts, viable vaccines have yet to materialize to control the diseases they are associated with including the African trypanosomes. While it is difficult to estimate the actual numbers of people who have contracted human sleeping sickness, as most of the transmission is occurring in war-torn countries and the impacted communities are often isolated by ongoing political instability, the numbers are thought to be in millions (Ekwanzala, Pepin et al. 1996; Moore, Richer et al. 1999).

The current management of human diseases relies on elimination of infected hosts by chemotherapy (McNeil 2000) while the management of animal diseases largely relies on vector control strategies, including ground and aerial spraying as well as application of insecticides directly to cattle in farming communities. Recently, traps and targets have also been employed, but reports on their sustainability have been mixed (Gouteux and Sinda 1990). It is more likely that a combination of several methods in an integrated, phased and area-wide approach would be more effective in controlling these diseases. Using such an integrated approach of population suppression followed by the Sterile Insect Technique (SIT), it has been possible to eradicate *Glossina austeni* from the island of Zanzibar (Vreysen, Saleh et al. 2000). The Organization of African Unity (OAU) has reached a consensus decision to eradicate tsetse flies from Africa at their Summit Meeting in Togo in July 2000, (Decision No: AHG/Dec. 156 (XXXVI) using such area-wide approaches. This is important because tsetse flies occupy vast regions, and any satisfactory control program will require the participation of all countries affected.

The recent advances in recombinant DNA technologies and their application to molecular genetic approaches for the control of vector-borne-diseases can provide alternative avenues but can also enhance the existing area-wide strategies by improving their efficacy and reducing their cost (Aksoy, Maudlin et al. 2001). The most challenging application of these technologies is transgenesis, which aims to genetically manipulate the ability of insects to transmit pathogens; i.e. modulate their vector competence. For this approach to succeed, it is necessary to develop a genetic transformation system so that foreign DNA can be reliably introduced and expressed in insects in tissues where the products can interact with parasites, genes encoding for anti-trypanosomal products need to be identified and methodologies are needed to drive the engineered insects into field populations to reduce disease. Since the same tsetse species are involved in the transmission of both the human and animal disease causing trypanosomes, the vector-based technologies proposed here stand to benefit the control of both diseases.

TRANSGENESIS

At the core of transgenesis is the process of genetic transformation, which in many insects relies on the microinjection of transposable elements into eggs to mediate germ-line transformation. It has now been possible to introduce foreign genes into several important insect vectors including one important malaria vector in Asia, *Anopheles stephensi* (Catteruccia, Nolan et al. 2000). Using gene specific expression systems, anti-parasitic gene products can then be expressed in tissues which are involved in parasite transmission, such as the gut, salivary glands or fatbody.

Recently, an antimicrobial immune molecule, defensin, has been successfully expressed in the hemolymph of transgenic *Aedes agypti* mosquitoes using the transcription elements of a blood-meal induced fatbody tissue-specific gene (Kokoza, Ahmed et al. 2000). Tsetse flies, however, have an unusual reproductive biology as there is no free egg stage, the females retain each egg within the uterus. Following fertilization and hatching, one young larva matures and is expelled as a fully developed third instar larva. Each female can deposit 2-3 offspring during its 5-6 week life span in the field. This viviparous reproductive biology complicates attempts to transform tsetse through egg microinjection. Tsetse, however, harbor bacterial symbiotic organisms which can be exploited to express foreign genes (Aksoy 2000). Since these bacteria live in close proximity to trypanosomes in the gut, anti-pathogenic products expressed in symbionts could adversely effect trypanosome transmission.

TSETSE SYMBIONTS

Many insects with limited diets such as blood, plant sap or wood rely on symbiotic microorganisms to fulfill their nutritional requirements (Buchner 1965). Because it has not been possible to cultivate most of these fastidious and often intracellular organisms *in vitro*, their correct taxonomic positioning has been controversial. Recent advances in PCR-based technologies as well as the use of nucleic acid sequences in phylogenetic reconstruction have now provided insight into their relationships.

Microorganisms with different microscopic characteristics have also been reported from various tissues of tsetse including midgut, hemolymph, fat body and ovaries (reviewed in Aksoy 2000). Phylogenetic studies have now shown that they represent three different microbes. The obligate primary symbiont (genus *Wigglesworthia*) (Aksoy 1995; Aksoy, Pourhosseini et al. 1995) and the mutualist secondary symbiont (genus *Sodalis*) (Aksoy, Pourhosseini et al. 1995; Cheng and Aksoy 1999; Dale and Maudlin 1999) are enteric and are closely related to *Escherichia coli,* while the third parasitic symbiont (genus *Wolbachia*) is a member of Rickettsiaceaea (O'Neill, Gooding et al. 1993).

Using a bacterium-specific PCR-amplification assay, *Wigglesworthia* is found to live within specialized epithelial cells in the bacteriome-tissue in anterior gut while *Sodalis* lives both inter- and intracellularly in midgut, muscle, fatbody, hemolymph, milkgland

and in certain species in salivary gland tissues (Cheng and Aksoy 1999). When tsetse and *Wigglesworthia* phylogenies were independently determined and compared, they were found to display identical relationships among the different species. This finding suggests that a tsetse ancestor was infected with a bacterium, and from this ancestral pair evolved the tsetse host and endosymbiont, comprising the species of tsetse and *Wigglesworthia* strains that exist today and overruling possible horizontal transfer events between species. Molecular clock analysis indicates that this symbiosis is ancient and might have been established about 80-100 million years ago (Chen, Song et al. 1999). Similar characterization of *Sodalis* has indicated that this association is more recent and may represent horizontal transfer events between species (Chen, Song et al. 1999). The phylogenetic analysis of *Wolbachia* strain types infecting different species of tsetse has shown that each are different and represent independent acquisitions (Zhou, Rousset et al. 1998; Cheng, Ruel et al. 2000).

It has been difficult to study the individual functions of the multiple symbionts in tsetse as attempts to eliminate the symbionts have been found to result in retarded growth of the insect and a decrease in egg production, preventing the ability of the aposymbiotic host to reproduce (Nogge 1976). This inability to reproduce can be partially restored when the aposymbiotic tsetse flies receive a bloodmeal that is supplemented with B-complex vitamins (thiamine, pantothenic acid, pyridoxine, folic acid and biotin) suggesting that the endosymbionts probably play a role in metabolism that involves these compounds (Nogge 1981). During its intrauterine life, the tsetse larva receives nutrients along with both gut symbionts from its mother via milk-gland secretions (Ma and Denlinger 1974; Aksoy, Chen et al. 1997) while *Wolbachia* is transmitted transovarially. Given the reproductive biology of tsetse, all three symbionts are maternally transmitted to the progeny.

TRYPANOSOME TRANSMISSION IN TSETSE

The successful transmission of the *T. brucei* complex parasites involves two developmental stages in the tsetse host; first, differentiation of the ingested mammalian form parasites to insect-stage procyclic cells in midgut and proventriculus tissues; and second, the invasion and maturation of these to metacyclic forms in salivary glands. This developmental cycle requires 20-30 days before the parasites can be transmitted to another host via the tsetse bite. The *T. congolense* and *T. vivax* trypanosomes complete their differentiation in the gut, hypopharynx and proboscis tissues of the fly. The molecular and biochemical events resulting in the successful transmission of a trypanosome to its mammalian host are complex and apparently involve tsetse, their symbionts and trypanosome derived factors (Welburn and Maudlin 1999). The characterization of these mechanisms will undoubtedly result in a greater degree of insight into host parasite interactions and may also help to identify potential targets that can be used to block disease transmission in the fly. However, although our knowledge of these pathways is incomplete, transgenic technology gives us the potential to interfere with parasite transmission by expressing foreign genes with antitrypanosomal properties *in vivo* in tsetse.

EXPRESSION OF FOREIGN GENES IN TSETSE SYMBIONTS

It has been possible to establish an *in vitro* culture of *Sodalis* (Welburn, Maudlin et al. 1987; Beard, O'Neill et al. 1993). The availability of this *in vitro* culture has allowed us to develop a genetic transformation system for this organism using the broad host range replicon *oriV* derived from a *Pseudomonas aeruginosa* plasmid (Beard, O'Neill et al. 1993). A foreign gene product, green fluorescent protein (GFP) has also been expressed in *Sodalis* and the *in vitro* manipulated recombinant *Sodalis* has been found to be successfully acquired by the intrauterine progeny when microinjected into the female parent hemolymph. The recombinant symbionts have been shown to be transmitted to F1 and F2 flies and successfully express the GFP marker gene product (Cheng and Aksoy 1999). Since *Sodalis* is found in the gut in close proximity to where trypanosomes differentiate and replicate, the expression and secretion of antitrypanosomal gene products in the recombinant symbionts *in vivo* could disrupt parasite differentiation or establishment in the gut. In addition, as *Sodalis* is present in various somatic tissues including the hemolymph and in some species salivary glands, expressed foreign gene products can impact trypanosome development in various compartments.

Once a genetic transformation approach is developed, anti-trypanosomal molecules must be identified to induce a parasite refractory phenotype in tsetse. Among the potential products are various immune modulators such as the antimicrobial peptides defensin and cecropin, which have been shown to have antitrypanosomal effects (Beard, Mason et al. 1992). Using a similar symbiont-based insect transformation approach, it has been possible to block the transmission of *Trypanosoma cruzi* in *Rhodnius prolixus in vivo* by expressing cecropinA (cecA) in its symbiont, *Rhodococcus rhodnii* in the hindgut of the bugs (Durvasala, Gumbs et al. 1997).

The identification of monoclonal antibodies (mAbs) with parasite transmission blocking characteristics and their expression as single-chain antibody gene fragments in the symbionts provides a vast array of potential antipathogenic products. Recently, it has been possible to express and secrete a single-chain antibody gene product in the transformed symbionts of reduviid bugs *in vivo* (Durvasula, Gumbs et al. 1999). Towards this end, several transmission-blocking antibodies targeting the major surface protein of the insect stage procyclic trypanosomes have also been reported (Nantulya and Moloo 1988).

The relative ease of transformation and gene expression in bacteria, and the multitude of potential antiparasitic targets that can be explored make this a highly desirable system for transgenic approaches. Should resistance develop in parasites against the expressed foreign gene product, it would be relatively easy to express a different product. Alternatively, several target genes can potentially be expressed simultaneously in the symbionts to prevent the development of resistance against any one individual target.

FIELD APPLICATIONS OF TRANSGENESIS

In order to interfere with disease transmission, the naturally susceptible field population needs to be replaced with their engineered refractory counterparts. A powerful potential driving system involves the use of *Wolbachia* symbionts. The functions of *Wolbachia* in the various insect hosts they infect are variable. However, one common reproductive abnormality that *Wolbachia* induces in its host has been termed cytoplasmic incompatibility (CI), and when expressed results in embryonic death due to disruptions in early fertilization events (Bandi C, Dunn et al. 2000).

In an incompatible cross, the sperm enters the egg but does not successfully contribute its genetic material to the potential zygote. In most species, this results in very few hatching eggs. The infected females have a reproductive advantage over their uninfected counterparts as they can produce progeny after mating with both the imprinted (*Wolbachia* infected male) and normal sperm. This reproductive advantage allows *Wolbachia* to spread into populations. In *D. simulans* in the central California valley, a natural *Wolbachia* infection invading naïve populations has spread at a rate of over 100 km per year simply through the expression of CI (Turelli and Hoffmann 1991). Different tsetse populations surveyed from the field were found to harbor distinct *Wolbachia* infections and the prevalence of infections in tsetse field populations has shown that significant polymorphism exists in the field (Cheng, Ruel et al. 2000).

In order to understand the functional role of *Wolbachia* in insects, it has been possible to cure most insects of their *Wolbachia* infections by administering antibiotics in their diet. This approach, however, has not been feasible in tsetse since the antibiotic treatment results in the clearing of all symbionts, including the obligate symbiont *Wigglesworthia*, and in the absence of these, the flies become sterile. Hence, in order to study the functional significance of *Wolbachia* in tsetse, uninfected flies need to be collected from the polymorphic field populations and colonized so that appropriate mating experiments can be done in the laboratory.

As *Wolbachia* infected insects replace naïve populations by virtue of the CI phenomena, they can drive other maternally inherited elements into that same population such as mitochondria or other maternally inherited organisms such as the gut-symbionts of tsetse (Beard, O'Neill et al. 1993). It has been proposed that multiple *Wolbachia* infections in which an insect contains two or more different *Wolbachia* strains that are incompatible with each other could be used to generate repeated population replacements or to spread *Wolbachia* into target species that already contain an existing infection (Sinkins, Braig et al. 1995; Sinkins, Curtis et al. 1997). Also the experimental transfer of *Wolbachia* between different hosts and even into insects with no prior infection history is now possible (reviewed in Sinkins 1997).

The use of engineered *Sodalis* to generate trypanosome refractory tsetse flies and the potential use of *Wolbachia* symbionts to drive these phenotypes in nature are schematically depicted below.

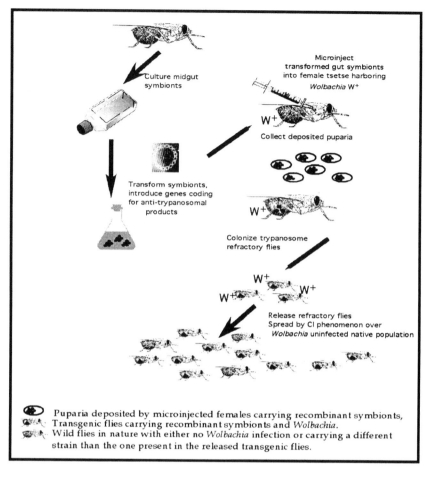

Puparia deposited by microinjected females carrying recombinant symbionts,
Transgenic flies carrying recombinant symbionts and *Wolbachia*.
Wild flies in nature with either no *Wolbachia* infection or carrying a different
strain than the one present in the released transgenic flies.

TRANSGENIC APPROACHES AND SIT

Transgenic approaches also stand to improve the existing vector control strategies, including SIT which has been successfully used to eradicate tsetse flies from the island of Zanzibar (Vreysen, Saleh et al. 2000). For SIT, the insects are mass reared in large-scale insectaries, the males are sterilized by irradiation and then taken to the selected area and released by air. The sterile males fertilize wild females, which then are unable to produce progeny. By continually releasing sterile males over a period of 3-4 generations, the target population can be eradicated (Vreysen, Saleh et al. 2000). Improvements in two aspects of current tsetse SIT technology have the potential to enhance its efficacy. The first is the use of *Wolbachia* mediated CI as a method of inducing sterility as an alternative to irradiation. In this approach, the release strain of tsetse would carry a *Wolbachia* infection that would induce CI when males are mated with wild females. As the competitiveness of these males would be expected to be much higher than irradiated males, fewer insects would need to be released in order to achieve the same level of sterility in the wild population. This strategy is dependent on the use of a very efficient sexing system. If *Wolbachia*-infected females are released in sufficient quantities, then *Wolbachia* will have the opportunity to invade the target population, which would render subsequent releases ineffective. This approach has been successfully tested in *Culex* mosquitoes (Shahid and Curtis 1987). If it was impossible to guarantee extremely low quantities of released females then it would be possible to incorporate low levels of irradiation with *Wolbachia* induced sterility to prevent released females from successfully reproducing. The second is the development of tsetse strains which are refractory for trypanosome transmission since the male flies released in large numbers can potentially contribute to a temporary increase in disease transmission. The refractoriness traits conferred by the recombinant *Sodalis* can be incorporated directly into the release strains for SIT. The use of sterile transgenic insects removes the possibility of vertical transmission of the transgene.

PERSPECTIVES

The morbidity and mortality associated with vector-borne diseases are overwhelming not to mention the social and economic burdens these diseases impose in the developing world. Much effort has gone into the development of control strategies to combat these pathogens in their mammalian hosts. In many cases, however the pathogens were found to be armed with various mechanisms which allowed them to bypass these conventional approaches such as vaccines and drug therapy. Similarly, many insects have now developed resistance to insecticides which were extensively used for control of these pests. Hence, innovative strategies are urgently needed to combat these diseases. It is hoped that some of these innovative strategies will arise from the ongoing parasite and vector genome projects. On the vector side, the application of transgenesis to block pathogen transmission in insects will undoubtedly be realized in the laboratory. The successful application of these technologies in the field will be more difficult. The concerns focused on various ethical, safety, regulatory and political issues will need to be discussed and addressed in light of the benefits and risks associated with these technologies. Towards this end, much progress is being made in the application of similar technologies for control of agricultural pests. In the meantime, the rapid advances being made in the molecular genetics of vector disease research could provide immediate benefits by being able to augment the more conventional and ongoing control approaches.

REFERENCES

Aksoy, S. (1995). "*Wigglesworthia* gen. nov. and *Wigglesworthia glossinidia* sp. nov., taxa consisting of the mycetocyte-associated, primary endosymbionts of tsetse flies." Int J Syst Bacteriol **45**(4): 848-51.

Aksoy, S. (2000). "Tsetse: a haven for microorganisms." Parasitology Today **16**(3): 114-119.

Aksoy, S., X. Chen, et al. (1997). "Phylogeny and potential transmission routes of midgut-associated endosymbionts of tsetse (Diptera:Glossinidae)." Insect Mol Biol **6**(2): 183-90.

Aksoy, S., I. Maudlin, et al. (2001). "Prospects for control of African trypanosomiasis by tsetse vector manipulation." Trends in Parasitology **17**(1): 29-35.

Aksoy, S., A. A. Pourhosseini, et al. (1995). "Mycetome endosymbionts of tsetse flies constitute a distinct lineage related to *Enterobacteriaceae*." Insect Mol Biol **4**(1): 15-22.

Bandi C, A. Dunn, et al. (2000). "Inherited Microorganims, Sex-specific Virulence and Reproductive Parasitism." Parasitology Today *in press*.

Beard, C., P. Mason, et al. (1992). "Transformation of an insect symbiont and expression of a foreign gene in the Chagas' disease vector *Rhodnius prolixus*." Am J Trop Med Hyg **46**(2): 195-200.

Beard, C., S. O'Neill, et al. (1993). "Modification of arthropod vector competence via symbiotic bacteria." Parasitology Today **9**(5): 179-183.

Beard, C. B., S. L. O'Neill, et al. (1993). "Genetic transformation and phylogeny of bacterial symbionts from tsetse." Insect Mol Biol **1**(3): 123-31.

Buchner, P. (1965). Endosymbiosis of Animals with Plant Micro-organisms. New York, Interscience Publishers Inc.: 210-338.

Catteruccia, F., T. Nolan, et al. (2000). "Stable germline transformation of the malaria mosquito *Anopheles stephensi*." Nature **405**(6789): 959-62.

Chen, X., L. Song, et al. (1999). "Concordant evolution of a symbiont with its host insect species : Molecular phylogeny of genus *Glossina* and its bacteriome-associated endosymbiont, *Wigglesworthia glossinidia*." Journal of Molecular Evolution **48**(1): 49-58.

Cheng, Q. and S. Aksoy (1999). "Tissue tropism, transmission and expression of foreign genes *in vivo* in midgut symbionts of tsetse flies." Insect Molecular Biology **8**(1): 125-132.

Cheng, Q., T. Ruel, et al. (2000). "Tissue distribution and prevalence of *Wolbachia* infections in tsetse flies, *Glossina* spp." Medical and Veterinary Entomology **14**(1): 51-55.

Dale, C. and I. Maudlin (1999). "*Sodalis* gen. nov. and *Sodalis glossinidius* sp. nov., a microaerophilic secondary endosymbiont of the tsetse fly *Glossina morsitans morsitans*." Int J Sys Bacteriology **49**(1): 267-275.

Durvasala, R., A. Gumbs, et al. (1997). "Prevention of insect borne disease: an approach using transgenic symbiotic bacteria." Proceedings National Academy of Sciences USA **94**: 3274-3278.

Durvasula, R., A. Gumbs, et al. (1999). "Expression of a functional antibody fragment in the gut of *Rhodnius prolixus* via transgenic bacterial symbiont *Rhodococcus rhodnii*." Med Vet Entomology **13**(2): 115-119.

Ekwanzala, M., J. Pepin, et al. (1996). "In the heart of darkness:sleeping sickness in Zaire." Lancet **348**: 1427-1430.

Gouteux, J. and D. Sinda (1990). "Community participation in the control of tsetse flies. Large scale trials using the pyramid trap in the Congo." Trop Med Parasitology **41**(1): 49-55.

Kokoza, V., A. Ahmed, et al. (2000). "Engineering blood meal-activated systemic immunity in the yellow fever mosquito, Aedes aegypti." Proc Natl Acad Sci U S A **97**(16): 9144-9.

Ma, W.-C. and D. L. Denlinger (1974). "Secretory discharge and microflora of milk gland in tsetse flies." Nature **247**: 301-303.

McNeil, D. (2000). Drug Companies and Third World: A case study in neglect. New York Times. NY: 1.

Moore, A., M. Richer, et al. (1999). "Resurgence of sleeping sickness in Tambura County, Sudan." Am J Trop Med Hyg **61**(2): 315-8.

Nantulya, V. M. and S. K. Moloo (1988). "Suppression of cyclical development of *Trypanosoma brucei brucei* in *Glossina morsitans centralis* by an anti-procyclics monoclonal antibody." Acta Trop **45**(2): 137-44.

Nogge, G. (1976). "Sterility in tsetse flies (*Glossina morsitans* Westwood) caused by loss of symbionts." Experientia **32**(8): 995-6.

Nogge, G. (1981). "Significance of symbionts for the maintenance of an optional nutritional state for successful reproduction in hematophagous arthropods." Parasitology **82**: 101-104.

O'Neill, S. L., R. H. Gooding, et al. (1993). "Phylogenetically distant symbiotic microorganisms reside in *Glossina* midgut and ovary tissues." Med Vet Entomol 7(4): 377-83.

Shahid, M. A. and C. F. Curtis (1987). "Radiation sterilization and cytoplasmic incompatibility in a "tropicalized" strain of the Culex pipiens complex (Diptera: Culicidae)." J Med Entomol **24**(2): 273-4.

Sinkins, S. P., H. R. Braig, et al. (1995). "*Wolbachia* superinfections and the expression of cytoplasmic incompatibility." Proc R Soc Lond B Biol Sci **261**(1362): 325-30.

Sinkins, S. P., C. F. Curtis, et al. (1997). The potential application of inherited symbiont systems to pest control. Influential Passengers. S. L. O'Neill, A. A. Hoffmann and J. H. Werren. Oxford, Oxford University Press: 155-175.

Turelli, M. and A. A. Hoffmann (1991). "Rapid spread of an inherited incompatibility factor in California Drosophila." Nature **353**(6343): 440-2.

Vreysen, M. J., K. M. Saleh, et al. (2000). "*Glossina austeni* (Diptera: Glossinidae) eradicated on the Island of Unguga, Zanzibar, using the sterile insect technique." J. Econ. Entomology **93**: 123-135.

Welburn, S. and I. Maudlin (1999). "Tsetse-trypanosome interactions: Rites of Passage." Parasitology Today **15**(10): 399-403.

Welburn, S. C., I. Maudlin, et al. (1987). "In vitro cultivation of rickettsia-like-organisms from Glossina spp." Ann Trop Med Parasitol **81**(3): 331-5.

Zhou, W., F. Rousset, et al. (1998). "Phylogeny and PCR-based classification of *Wolbachia* strains using WSP gene sequences." Proc Royal Society London Ser B Biological Sciences **265**: 509-515.

DIAGNOSIS OF HUMAN AND ANIMAL AFRICAN TRYPANOSOMIASIS

P. Büscher
Department of Parasitology, Institute of Tropical Medicine Antwerp, Belgium

ABSTRACT

Diagnosis of African trypanosomiasis has been the subject of intensive research for decades, which is reflected by the bulk of publications related to this topic. From this literature, it appears that the current parasite detection techniques can hardly be improved and that reliable antigen detection tests will remain wishful thinking. Much more success has been obtained in the development of antibody detection tests of which some have even reached the end-user, *i.e.* health personnel and veterinarians in rural regions of Africa, Latin America and Asia. Unfortunately, the production of diagnostics for African trypanosomiasis has no economical value. As a consequence, potentially excellent diagnostic tests never reach the validation phase, not to mention production, and sooner or later will belong to history unless they have some scientific value. In this context, international human and animal health organizations have largely missed the opportunities to provide validated diagnostics to those who need them. Recent literature clearly reflects the general shift from serological diagnostics to molecular diagnostics. Molecular diagnostics have unequalled potentials for sensitive and specific detection of human and animal African trypanosomiasis. However, the risk that molecular diagnostics will remain the property of the scientific world with only minor impact on real disease control should not be underestimated.

Key words: Diagnosis, African trypanosomiasis, surra, nagana, dourine, sleeping sickness, *Trypanosoma brucei*, *Trypanosoma congolense*, *Trypanosoma vivax*, *Trypanosoma evansi*, *Trypanosoma equiperdum*.

INTRODUCTION

African trypanosomiasis is a general term for infections in many different hosts (man, bovine, buffalo, goat, sheep, camel, horse, pig and wild animals) caused by various trypanosome species (*Trypanosoma brucei, T. congolense, T. vivax, T. evansi, T.*

equiperdum) and subspecies. As a consequence, a chapter like this must remain superficial. Nevertheless, some interesting general features on the development of African trypanosomiasis diagnostics can be observed.

Despite decades of research into the development of reliable diagnostic tests for African trypanosomiasis, little has changed in current practice, particularly at the level of farmers, veterinarians and health personnel in developing countries where African trypanosomiasis prevails. In fact, diagnosis of African trypanosomiasis in animals is too often based on clinical suspicion only, whereafter individual or herd treatment measures are undertaken. Only in human African trypanosomiasis (sleeping sickness), is parasitological confirmation of clinical or serological suspicion compulsory, because treatment is expensive and not without risk for the patient.

Most probably, the reason why simple and reliable diagnostic field tests are still rare must be found in the combination of limited economical interest and the inconstancy of research priorities and related financial resources. It is amazing to see how many publications describe new or improved diagnostic techniques, mostly evaluated on a restricted, well-controlled set of samples and presenting promising 'preliminary' results. Some of them are indeed promising and based on sound scientific research. Unfortunately, competition for the same restricted funding resources and the constant pressure to come up with something 'new', whether or not it will ever be applicable, results in a lack of encouragement for researchers to undertake multicenter validation of diagnostic tests that could then become applied on a large scale. Indeed, the researcher may encounter actual barriers against undertaking multicenter validation of diagnostic tests. It might be interesting to evaluate some day the role and activities of international organizations like WHO, FAO, ISCTRC, EU etc. in the light of the overt failure to really improve basic diagnosis of African trypanosomiasis.

Only for epidemiology and surveillance purposes, where perfect performance, simplicity and low cost are not too important, have some major improvements in African trypanosomiasis diagnosis been achieved with recent techniques such as ELISA and PCR. However, also in this field, standardization is definitely lacking. With few exceptions, most research groups prefer to develop their own system which is then applied in their action region with minor attention to multicenter validation of the tests. Faced with increasing drug resistance, expansion of the distribution area of some trypanosome species and their re-appearance in 'controlled' regions, one might expect researchers to focus onto the validation and standardized quality production of diagnostic tests.

Students who have no experience with African trypanosomiasis diagnosis may not be aware of the intrinsic difficulties to be overcome. Generally, clinical symptoms are rather unspecific or may be even obscure, particularly in trypanotolerant hosts. In many instances, the parasite load is extremely low which makes the detection of trypanosomes rather cumbersome and poorly sensitive. For example, with the most sensitive techniques, *T.b. gambiense*, the causative agent of sleeping sickness in West and Central Africa is probably detected in only 80% of the patients. The same applies to *T. evansi* in buffalo and camel or *T. congolense* and *T. vivax*, often occurring as mixed infections, in trypanotolerant cattle. The situation is even worse for *T. equiperdum* (dourine) in horses where parasite detection sensitivity is almost zero. Indirect tests based on antibody detection may be very sensitive but specificity may be poor due to cross-reaction with other infections, depending on the purity of the antigen applied. In addition, antibodies tend to remain detectable for months after cure, which makes antibody detection tests useless for follow-up studies. Finally, except for agglutination tests, for every host species, the appropriate conjugates should be incorporated in the test. This conjugate should preferentially be an anti-IgG to avoid cross-reactions due to aspecific IgM. In many instances, related to the low parasite load, the amount of circulating antigens is largely insufficient to detect by simple techniques. The same may apply for DNA detection tests such as PCR although these tests are theoretically very sensitive. The occurrence of mixed infections and of different taxa within each trypanosome species further complicates the picture. Special attention should be paid to choosing primers able to discriminate the described taxa, knowing that poorly studied taxa (e.g. within the *T. vivax* or *T.b. gambiense* group) may not be detected.

In the following paragraphs, an overview of some more or less recent achievements in the diagnosis of African trypanosomiasis is given.

PARASITE DETECTION

As stated above, parasite detection is cumbersome in the many cases where only low numbers of trypanosomes circulate in the host body fluids. Techniques for concentration of the trypanosomes by centrifugation of a blood sample are still the most widely applied in animal trypanosomiasis (*T. congolense, T. vivax, T. brucei, T. evansi*). After centrifugation of some blood in a capillary tube, the trypanosomes can be detected directly under the microscope at the level of the white blood cell layer, the buffy coat (Woo 1969). Where differential diagnosis is needed, the capillary tube is broken and the buffy coat is spread on the microscope slide for examination according to Murray et al. (1977). It is clear that the latter technique is not without risk of infection when human samples are manipulated. Our experience has also shown us that the much bigger and very motile microfilaria, often present in the blood of people living in sleeping sickness areas, strongly interferes with trypanosome detection. A more sensitive technique is the mini Anion Exchange Centrifugation Technique, developed by Lumsden et al. (1979). The mAECT is based on the separation of trypanosomes from the host blood through an anion exchange gel. The technique works best with the three subspecies of *T. brucei* and with *T. evansi*. The elution buffer may need adaptation to the host species. Some improvements to the original techniques using modern material have been published (Zillmann et al. 1996). For sleeping sickness diagnosis, a concentration technique originally developed for malaria, the Quantitative Buffy Coat (Bailey et al. 1992) has been introduced and is now applied by those who can pay for it. The technique combines facilitated visualization of trypanosomes by their concentration at the expanded buffy coat level, their motility and the staining of their nucleus and kinetoplast by acridine orange. Detection of trypanosomes in fresh thin blood preparations using this dye has been proposed long ago (Carrie1981) but has never been applied in the field. A major drawback of acridine orange is the simultaneous staining of the white blood cell nuclei. Therefore, we are currently studying techniques for species specific fluorescent staining of the trypanosomes in (fixed) blood samples by monoclonal antibodies. However, the small amounts of examined blood will always limit the sensitivity of such tests. Techniques based on expansion of trypanosomes through culture in laboratory animals or *in vitro* exist but their value is limited to experimental studies. It is generally accepted that not much progress can be expected in parasite detection techniques, at least not for direct diagnosis. This has serious consequences on the evaluation of indirect diagnostic tests due to the lack of a 'gold standard'.

ANTIBODY DETECTION

Indirect evidence of infection with African trypanosomes can be obtained through the presence of specific circulating antibodies. Sometimes, it is of interest to combine antibody detection with hematocrit estimation. Particularly in trypanosusceptible animals such

as camels and some breeds of ruminants, low PCV values together
with the presence of specific antibodies are highly predictive for
active infection. A wealth of different antibody detection systems
exists ranging from direct and indirect agglutination,
immunofluorescent assays and ELISA. They are adapted for testing
blood (fresh or dried on filter paper), serum, plasma and cerebrospinal
fluid from all kind of host species. They are extremely useful to study
the prevalence of African trypanosomiasis in a population and
particularly in *T.b. gambiense* sleeping sickness, the introduction of a
simple, quick direct agglutination test, the CATT (Magnus et al.
1978) has been a major breakthrough in the control of this disease by
selecting the seropositive individuals for parasite examination to
confirm the infection. The test is based on detection of antibodies
against predominant surface antigens of *T.b. gambiense*. A similar
test exists for *T. evansi* (CATT/*T.evansi* (Bajyana Songa et al. 1988).
Unfortunately, the antigenic variation of *T.b. brucei* and *T.b.
rhodesiense* is too large to allow the development of CATT tests for
these subspecies. For the latter subspecies, another direct
agglutination test exists (Liu et al. 1989) which recently has been
simplified by using fixed instead of live procyclic trypanosomes
(Ngaira et al. 1999). Knowledge about the antigenic repertoires of *T.
congolense* and *T. vivax* is almost non-existent. Antibody detection
tests for these species make use of more or less purified but still crude
antigens leaving room for aspecific cross-reactions. The situation is
less problematic for *T. congolense* than for *T. vivax* and *T. brucei*. In
regions where *T. vivax* and *T. brucei* or *T. evansi* occur as mixed
infections it is almost impossible to make the distinction at the level
of circulating antibodies. As mentioned earlier, almost every research
or surveillance group produces its own antigens on a day by day basis
leaving us with poorly standardized tests. A remarkable exception is
the effort of IAEA, Vienna, to standardize an ELISA test for *T. vivax*
and *T. congolense* in bovine and to have it validated and introduced in
many African laboratories (Rebeski et al. 1999b, 2000). This work is
a fine example of difficulties that may be encountered during
development of ELISAs for African trypanosomes ranging from the
choice of the microtiter plates, the standardized production of the
native antigens, the choice of appropriate controls and conjugates, the
stabilization of the test kit and the analysis of the data. Unfortunately,
the IAEA project came to an end and the continuation of the
production is not guaranteed, partly due to the low economical value.
It is interesting to observe that many valuable antigen candidates for
antibody detection have been described such as Invariable Surface
Glycoproteins (Ziegelbauer et al. 1992) and cytoskeleton proteins

from *T. brucei* (Imboden et al. 1995) and a cysteine protease from *T. congolense* (Authié et al. 1993) but that this not necessarily leads to the development of diagnostic tests. Reasons for this must be found in the fact that sufficient amounts of pure native antigen are difficult to prepare and that groups involved in fundamental research have neither the experience nor the interest to develop diagnostic tests. However, efforts to replace native antigens by recombinant ones in order to simplify production and to increase species specificity of antibody detection tests still continue e.g. at ILRI, Nairobi. These attempts, seldom published, are hindered by the well known polyclonal activation of B-cells in trypanosome infected hosts, leading to undesired cross-reactions when the recombinant antigen contains impurities. Probably only the use of synthetic peptides may lead to the desired sensitivity and specificity. In principle, such peptides may be applied in different test systems, as there are latex agglutination, ELISA, immunofluorescence or lateral flow devices. Considering the price of lateral flow devices it is difficult to believe that they ever will be produced for African trypanosomiasis, be it in animals or in man. One exception may be *T. equiperdum* by the mere fact that Europe and North America are threatened by its re-introduction. Only for *T. equiperdum*, our laboratory has been able to raise sufficient funding within the European context to start development of a new generation of antibody tests. *T. equiperdum* is believed to be the causative agent of dourine, a sexually transmitted disease in equidae. The trypanosome is very difficult to detect in the infected host. Therefore, antibody detection is used to screen for a possible infection. It appears that the only officially approved test by the OIE is the Complement Fixation Test, although it is generally accepted that this test cannot discriminate between *T. evansi* and *T. equiperdum*.

Development of antibody detection tests might seem relatively simple, their evaluation is far more difficult, particularly in the absence of a 'gold standard'. As long as a diagnostic test is evaluated on an experimentally infected population along with a naive control population, data are easy to analyze. However, for evaluation on naturally infected populations, other statistical techniques must be used such as the 'latent class analysis' or Bayesian methodology (EnŸe et al. 2000). This approach has not yet been followed in African trypanosomiasis diagnosis and should be encouraged by all means.

ANTIGEN DETECTION

For quite a long time, it was believed that antigen detection might overcome the problems encountered with parasite or antibody detection. Much research by different groups has been conducted on the development of antigen detection ELISAs (Liu and Pearson 1987; Nantulya and Lindqvist 1989; Olaho-Mukani et al. 1993) dipstick assays (Kashiwazaki et al. 1994) and latex agglutination (Nantulya 1994, 1997). Unfortunately, at present no antigen detection test with proven efficiency exists. Multiple explanations have been put forward such as the intrinsic low amount of circulating antigen which in addition may be bound to immune complexes, the nature of the capturing antibodies and their target antigen (Rebeski et al. 1999a) Other causes of failure could be the occurrence of anti-idiotypic antibodies that mimic the target antigen and circulating anti-mouse antibodies, generated by the polyclonal activation of B-cells, which may interfere with the antigen capturing. The existence of cryptic infections, which cannot be revealed by parasite detection techniques, has been postulated as well but for obvious reasons this cannot be tested. For those who are not convinced of the failure of antigen detection tests in African trypanosomiasis, it may be useful to consider specificity studies on animal or human samples from Europe where these infections are totally absent.

MOLECULAR DIAGNOSIS

Molecular diagnosis, mainly based on detection of trypanosome DNA, opens new perspectives for diagnosis of African trypanosomiasis when parasite detection fails. Studies on the genome of trypanosomes have led to the development of species and subspecies specific primers for PCR. To increase the sensitivity and the specificity, PCR is sometimes combined with hybridization (Clausen et al. 1999). Specific primers exist for *T. congolense* and *T. vivax* although for the latter species it is not certain that all strains are recognized by these primers. For the *Trypanozoon* group, the picture is less clear. Most of the primer sets cannot make the distinction between *T. evansi* and *T. brucei* s.l. nor between the subspecies of *T. brucei* as there are *T.b. brucei*, *T.b. gambiense* and *T.b. rhodesiense* (Pereira de Almeida 1999). In most instances, this is not a problem but when one is interested for example in the animal reservoir of human infective trypanosomes, PCR may not be useful until gambiense and rhodesiense specific primer sets are used. The fact that *T.b. gambiense* itself can be divided in at least two different groups further complicates the situation. For *T.b. gambiense* group I, specific primers already have been identified (Bromidge et al. 1993; Mathieu-Daudé et al. 1994; Biteau et al. 2000). Most probably, in the near future more and better primer sets will become available.

In principle, all types of biological host samples can be assayed for trypanosome DNA (blood, lymph, cerebrospinal fluid, tissue). A multitude of sample collection and DNA preparation methods exist ranging from the classical chloroform/phenol extraction through purification with the help of resins or silica. Many of these methods have already been tested in African trypanosomiasis without unexpected problems. However, it should be kept in mind that, depending on the technique, the risk of contamination during sample preparation under field conditions could be high.

In addition, it appears that the extreme low detection limit obtained with PCR on DNA directly purified from trypanosome cultures (less than 1 trypanosome per ml) is never obtained when starting with trypanosome containing blood from clinical samples. In those samples, the detection limit of PCR is comparable to that of the most sensitive parasitological tests (Kabiri et al. 1999; Omanwar et al. 1999). Furthermore, attention should be paid to the conservation of the samples. In our hands, the stabilization of blood samples in the AS1 buffer included in the Qiagen mini blood kit gave excellent results (Holland et al. 2000) while important loss of DNA was observed when plain blood samples are frozen or dried on filter paper. The availability of filter paper especially designed for DNA collection (e.g. FTA paper from Whatman) may bring the solution although the amount of blood that can be applied may be too small.

Apart from the fact that the current price and technical requirements of molecular diagnostics are prohibitive for their generalized application in developing countries, more and more evidence becomes available that also molecular tests suffer from unexplained false negative (Kabiri et al. 1999; Simo et al. 1999) and false positive results. PCR can remain positive long after successful cure (Kirchhoff 1998; Pereira de Almeida 1999) and almost half of a human control population in Côte d'Ivoire without any clinical, serological or parasitological evidence of infection was found positive in PCR (Garcia et al. 2000). Furthermore, repeated sampling and DNA extraction of the same individuals can yield contradictory results (Solano, personal communication). It is clear that still much work has to be done on the development and validation of molecular diagnostic tests for African trypanosomiasis starting with studies on the specificity of the existing primer sets against a large collection of trypanosome strains and other pathogenic organisms followed by specificity testing on clinical samples from non-endemic regions and sensitivity testing on confirmed infections. Again, the absence of a 'gold standard' should be taken into account and one should not be too confident that a PCR positive result is a definite proof of infection.

As parasitologists are natural optimists and can hardly wait for the validation of newly developed tests, investigations into the applicability of alternative molecular tests such as PCR-ELISA, multiplex PCR, PCR-RFLP and PNA have already been initiated. As mentioned in the introduction, international institutions may play a leading role in supporting multicenter studies with one major goal, which is the improvement of African trypanosomiasis diagnosis for the benefit of the end user.

STAGE DETERMINATION AND FOLLOW-UP IN SLEEPING SICKNESS

In sleeping sickness, the drug used for treatment depends on whether the parasite has reached the central nervous system. For treatment of the second or meningo-encephalitic stage, drugs that pass the blood-brain barrier in sufficient amounts should be used. Since there are no exclusive clinical signs indicating the evolution from the haemo-lymphatic to the meningo-encephalitic stage, the only way to determine the disease stage is by examination of the cerebrospinal fluid (CSF) obtained by lumbar puncture, assuming that the changes observed in the CSF reflect the events going on in the central nervous system (CNS). The same applies for follow-up after treatment: CSF should be obtained and examined regularly for up to two years before a patient can be considered cured. According to the recommendations of the WHO for stage determination and follow-up, the CSF has to be examined on white blood cell number, total protein concentration and presence of trypanosomes. Since there is no close relationship between these parameters, ideally all of them should be examined. Abnormal values, indicating central nervous system infection are as follows: > 5 cells/μl, presence of trypanosomes, protein concentration > 250 mg/l (trichloroacetic acid precipitation) or > 370 mg/l (dye binding method) or > 450 mg/l (sulfosalicylic acid precipitation). For people with some experience in the field, it is clear that the determination of these CSF parameters poses some problems: cell counting at the detection limit of the classical counting chambers suffers from low repeatability, 'normal' cell count strongly depends on age of the patient, elevated cell counts may be caused by other infections, 'normal' protein concentration depends on the method and the standard solution used, trypanosomes may be difficult to detect etc. In addition, little is known about the evolution of these parameters after treatment, except from the fact that it may take two years before the parameters have returned to normal. Rather recently, a limited number of research groups have renewed interest into the development of improved techniques for stage determination with some remarkable results.

For example, the replacement of classical cell counting chambers by disposable plastic counting chambers like the KOVA Glasstick slides from ICL retaining a fixed volume of CSF, greatly improves the reproducibility of the technique by reducing manipulation errors to a minimum (Lejon et al. 1998a). For detection of trypanosomes in the CSF, a modified single centrifugation technique (Miézan et al. 2000) has been proven to be much simpler and as sensitive as the double centrifugation technique. For more than 30 years, it is known that CSF of second stage sleeping sickness patients contains high amounts of IgM but for technical reasons it was impossible to quantify this IgM in African rural health centres. Only recently, a simple, rapid and stable agglutination test (LATEX/IgM) for IgM quantification in CSF has become available for application in the field (Lejon et al. 1998b). Detection of trypanosome specific antibodies in CSF has also become possible with a simple agglutination test, the LATEX/*Tb gambiense* (Büscher et al. 1999). For research purposes, an ELISA system has been developed for simultaneous measurement of trypanosome specific IgM and IgG in the CSF and in the serum thus allowing to calculate intrathecal antibody production as a marker for CNS infection (Lejon et al. 1999a). Other CSF markers for CNS inflammation and destruction are currently being studied such as anti-galactocerebroside antibodies (Bisser et al. 2000), neurofilament and glial fibrillary acidic protein (Lejon et al. 1999b) and cytokines (MacLean et al. 1999; Lejon et al. 1999c). The main interest of tests based on these markers probably lies in their potential to shorten the follow-up period after treatment.

Reports on the use of molecular diagnostic techniques for stage determination and follow-up are still rare (Kirchhoff 1998; Truc et al. 1999; Kyambadde et al. 2000)but application of these techniques on CSF samples is straightforward and may offer new tools, particularly to clinical treatment studies for cure rate assessment. Once again, prior validation of DNA detection techniques will be required.

REFERENCES

Authié E.G., G. Duvallet, C. Robertson and D.J.L. Williams. 1993. Antibody responses to a 33 kDa cysteine protease of Trypanosoma congolense: relationship to "trypanotolerance" in cattle. Parasite Immunology **15**: 465-474

Bailey J.W. and D.H. Smith. 1992. The use of the acridine orange QBC technique in the diagnosis of African trypanosomiasis. Transactions of the Royal Society of Tropical Medicine and Hygiene **86**: 630-630

Bajyana Songa E. and R. Hamers. 1988. A card agglutination test (CATT) for veterinary use based on an early VAT RoTat 1/2 of Trypanosoma evansi. Annales de la Société Belge de Médecine Tropicale **68**: 233-240

Bisser S., Z. Ayed, B. Bouteille, A. Stanghellini, J.C. Breton, M. Dumas and M.O. Jauberteau. 2000. Central nervous system involvement in African trypanosomiasis: presence of anti-galactocerebroside antibodies in patients' cerebrospinal fluid. Transactions of the Royal Society of Tropical Medicine and Hygiene **94**: 225-226

Biteau N., F. Bringaud, W. Gibson, P. Truc and T. Baltz. 2000. Characterization of Trypanozoon isolates using a repeated coding sequence and microsatellite markers. Molecular and Biochemical Parasitology **105**: 185-202

Bromidge T., W. Gibson, K. Hudson and P. Dukes. 1993. Identification of Trypanosoma brucei gambiense by PCR amplification of variant surface glycoprotein genes. Acta Tropica **53**: 107-119

Büscher P., V. Lejon, E. Magnus and N. Van Meirvenne. 1999. Improved latex agglutination test for detection of antibodies in serum and cerebrospinal fluid of Trypanosoma brucei gambiense infected patients. Acta Tropica **73**: 11-20

Carrie J. 1981. Colloration vitale fluorescente du trypanosome. Procédé rapide de détection du parasite dans le sang. *In* 26th meeting of the ISCTRC, Yaoundé, Cameroon, 1979. OAU/STRC (ed.) Eleza Services Ltd., Nairobi., p. 95-97.

Clausen P.-H., C. Waiswa, E. Katunguka-Rwakishaya, G. Schares, S. Steuber, and D. Mehlitz. 1999. Polymerase chain reaction and DNA probe hybridization to assess the efficacy of diminazene treatment in Trypanosoma brucei-infected cattle. Parasitology Research **85**: 206-211

EnŶe C., M.P. Georgiadis and W. O. Johnson. 2000. Estimation of sensitivity and specificity of diagnostic tests and disease prevalence when the true disease state is unknown. Preventive Veterinary Medicine **45**: 61-81

Garcia A., V. Jamonneau, E. Magnus, C. Laveissiere, V. Lejon, P. N'Guessan, L. N'Dri, N. Van Meirvenne and P. Büscher. Trop.Med.Int.Health, in press.

Holland W.G., F. Claes, L.N. My, N.G. Thanh, P.T. Tam, D. Verloo, P. Büscher, B. Goddeeris, and J. Vercruysse. International Journal of Parasitology, in press.

Imboden M.,N. Müller, A. Hemphill, R. Mattioli and T. Seebeck. 1995. Repetitive proteins from the flagellar cytoskeleton of African trypanosomes are diagnostically useful antigens. Parasitology **110**: 249-258

Kabiri M., J.R. Franco, P.P. Simarro, J.A. Ruiz, M. Sarsa, and D. Steverding. 1999. Detection of Trypanosoma brucei gambiense in sleeping sickness suspects by PCR amplification of expression-site-associated genes 6 and 7. Tropical Medicine and International Health **4**: 658-661

Kashiwazaki Y., K. Snowden, D.H. Smith and M. Hommel. 1994. A multiple antigen detection dipstick colloidal dye immunoassay for the field diagnosis of trypanosome infections in cattle. Veterinary Parasitology **55**: 57-69

Kirchhoff L.V. 1998. Use of a PCR assay for diagnosing African trypanosomiasis of the CNS: a case report. Central African Journal of Medicine **44**: 134-136

Kyambadde J.W., J.C.K. Enyaru, E. Matovu, M. Odiit and J.F. Carasco. 2000. Detection of trypanosomes in suspected sleeping sickness patients in Uganda using the polymerase chain reaction. Bulletin of the World Health Organization **78**: 119-124

Lejon V., P. Büscher and S. Bisser. 1998a. New tools for stage determination and follow-up in sleeping sickness. Antwerp, ITM. 40TH International Colloquium of the Institute of Tropical Medicine Antwerp "Sleeping Sickness Rediscovered", 14-18 December 1998.

-----------, P. Büscher, N.H. Sema, E. Magnus and N. Van Meirvenne. 1998b. Human African Trypanosomiasis: a latex agglutination field test for quantifying IgM in cerebrospinal fluid. Bulletin of the World Health Organization **76**: 553-558

-----------, P. Büscher, M.-P. Van Antwerpen, C. Sindic and F. Doua. 1999a. Quantitative and qualitative detection of intrathecal trypanosome specific antibody synthesis. *In* 25th Meeting of the ISCTRC, Mombasa 27th Sept.- 1st Oct. 1999. K. Sones (ed) OAU/STRC, Nairob.

-----------, J. Lardon, G. Kenis, S. Bisser, P. Büscher, E. Bosmans, and X. N'Siesi. 1999c. Cytokine concentrations in serum and cerebrospinal fluid of sleeping sickness patients. *In* 25th Meeting of the ISCTRC, Mombasa 27th Sept.- 1st Oct. 1999. K. Sones (ed) OAU/STRC, Nairobi

------------, L.E. Rosengren, P. Büscher, J.E. Karlsson and N.H. Sema. 1999b. Detection of light subunit neurofilament and glial fibrillary acidic protein in cerebrospinal fluid of Trypanosoma brucei gambiense -infected patients. American Journal of Tropical Medicine and Hygiene **60**: 94-98

Liu M.K., P. Cattand, I.C. Gardiner and T.W. Pearson. 1989. Immunodiagnosis of sleeping sickness due to Trypanosoma brucei gambiense by detection of antiprocyclic antibodies and trypanosome antigens in patients sera. Acta Tropica **46**: 257-266

----------- and T.W. Pearson. 1987. Detection of circulating trypanosomal antigens by double antibody ELISA using antibodies to procyclic trypanosomes. Parasitology **95**: 277-290

Lumsden W.H.R., C.D. Kimber, D.A. Evans and S.J. Doig. 1979. Trypanosoma brucei: miniature anion-exchange centrifugation technique for detection of low parasitaemias: adaptation for field use. Transactions of the Royal Society of Tropical Medicine and Hygiene **73**: 312-317

MacLean L., M. Odiit, D. Okitoi and J.M. Sternberg. 1999. Plasma nitrate and interferron-gamma in Trypanosoma brucei rhodesiense infections: evidence that nitric oxide production is induced during both early blood-stage and late meningoencephalitic-stage infections. Transactions of the Royal Society of Tropical Medicine and Hygiene **93**: 169-170

Magnus E., T. Vervoort and N. Van Meirvenne. 1978. A card-agglutination test with stained trypanosomes (C.A.T.T.) for the serological diagnosis of T.b.gambiense trypanosomiasis. Annales de la Société Belge de Médecine Tropicale **58**: 169-176

Mathieu-Daudé F., A. Bicart-See, M.-F. Bosseno, S.-F. Breniere and M. Tibayrenc. 1994. Identification of Trypanosoma brucei gambiense group I by a specific kinetoplast DNA probe. American Journal of Tropical Medicine and Hygiene **50**: 13-19

Miézan T.W., A.H. Meda, F. Doua, N.N. Djé, V. Lejon and P. Büscher. 2000. Single centrifugation of cerebrospinal fluid in a sealed pasteur pipette for simple, rapid and sensitive detection of trypanosomes. Transactions of the Royal Society of Tropical Medicine and Hygiene **94**: 293

Murray M., P.K. Murray and W.I.M. McIntyre. 1977. An improved parasitological technique for the diagnosis of African trypanosomiasis. Transactions of the Royal Society of Tropical Medicine and Hygiene **71**: 325-326

Nantulya V.M. 1994. Suratex(r): a simple latex agglutination antigen test for diagnosis of Trypanosoma evansi infections (Surra). Tropical Medicine and Parasitology **45**: 9-12

-----------. 1997. TrypTect CIATT: a card indirect agglutination trypanosomiasis test for diagnosis of Trypanosoma brucei gambiense and T. b. rhodesiense infections. Transactions of the Royal Society of Tropical Medicine and Hygiene **91**: 551-553

----------- and K.J. Lindqvist. 1989. Antigen-detection enzyme immunoassays for the diagnosis of Trypanosoma vivax, T. congolense and T. brucei infections in cattle. Tropical Medicine and Parasitology **40**: 267-272

Ngaira J.M., M.C. Wani, J.N. Njenga and S.O. Guya. 1999. The use of MOPATT, CIATT, PCR and parasitological tests in spot check screening of high-risk communities in selected villages where sleeping sickness has recently been reported in Teso District. *In* 25th Meeting of the ISCTRC, Mombasa 27th Sept.- 1st Oct. 1999. K. Sones (ed) OAU/STRC, Nairobi

Olaho-Mukani W., W.K. Munyua, M.W. Mutugi and A.R. Njogu. 1993. Comparison of antibody- and antigen-detection enzyme immunoassays for the diagnosis of Trypanosoma evansi infections in camels. Veterinary Parasitology **45**: 231-240

Omanwar S., J.R. Rao, S.H. Basagoudanavar, R.K. Singh and G. Butchaiah. 1999. Direct and sensitive detection of Trypanosoma evansi by polymerase chain reaction. Acta Vet. Hung. **47**: 351-359

Pereira de Almeida P.J.L. 1999. Contributions to the diagnostic evaluation of the Polymerase Chain Reaction for the detection of Salivarian trypanosomes. 1-195, Université Libre de Bruxelles, Brussels, Belgium.

Rebeski D.E., E.M. Winger, E.M.A. Van Rooij, R. Schöchl, W. Schuller, R.H. Dwinger, J.R. Crowther and P. Wright. 1999a. Pitfals in the application of enzyme-linked immunoassays for the detection of circulating trypanosomal antigens in serum samples. Parasitology Research **85**: 550-556

-----------, E.M. Winger, M.M. Robinson, C.M.G. Gabler, J.R. Dwinger and J.R. Crowther. 2000. Evaluation of antigen-coating procedures of enzyme-linked immunosorbent assay method for detection of trypanosomal antibodies. Veterinary Parasitology **90**: 1-13

-----------, E.M. Winger, Y-K. Shin, M. Lelenta, M.M. Robinson, R. Varecka and J.R. Crowther. 1999b. Identification of unacceptable background caused by non-specific protein adsorption to the plastic surface of 96-well immunoassay plates using a standardized enzyme-linked immunosorbent assay procedure. Journal of Immunological Methods **226**: 85-92

Simo G., P. Grebaut, S. Herder, S.W. Nkinin and L. Penchenier. 1999. Intérêt de la PCR dans le diagnostic de la trypanosomiase humaine africaine. Bulletin de liaison et de documentation OCEAC **32**: 17-21

Truc P., V. Jamonneau, G. Cuny and J.L. Frézil. 1999. Use of polymerase chain reaction in human African trypanosomiasis stage determination and follow-up. Bulletin of the World Health Organization **77**: 745-748

Woo P.T.K. 1969. The haematocrit centrifuge for the detection of trypanosomes in blood. Canadian Journal of Zoology **47**: 921-923

Ziegelbauer, K. and P. Overath. 1992. Identification of invariant surface glycoproteins in the bloodstream stage of Trypanosoma brucei. Journal of Biological Chemistry **267**: 10791-10796

Zillmann U., S.M. Konstantinov, M.R. Berger and R. Braun. 1996. Improved performance of the anion-exchange centrifugation technique for studies with human infective African trypanosomes. Acta Tropica **62**: 183-187

CHEMOTHERAPY OF AFRICAN TRYPANOSOMIASIS

John Richard Seed[1] and David W. Boykin[2], [1] Department of Epidemiology, University of North Carolina, Chapel Hill, NC, 27599; [2] Department of Chemistry, Georgia State University, Atlanta, GA 30303

ABSTRACT There is a critical need for new effective chemotherapeutic agents against the African Trypanosomes. At the present time, drug research is primarily focused on the molecular target approach, however, it is suggested that a more balanced approach involving screening against the whole organism be given greater emphasis. To illustrate this point, recent research involving the screening of new compounds derived from pentamidine is discussed. Analogues with improved chemotherapeutic potential have been obtained. Based upon the chemical structures of these lead compounds and a mode of action hypothesis the synthesis of novel trypanocidal drugs are in progress. Current data suggests that pentamidine analogues with oral bioavailability, as well as better blood-brain barrier transport have been developed. It is suggested that enlightened whole organism screening of analogues of the currently used chemotherapeutic agents will lead to the more rapid development of new trypanocidal drugs.

Key words Trypanosome, chemotherapy, new pentamidine analogues

INTRODUCTION

At the turn of the 20th century there were estimated to be over one million individuals infected with the African trypanosomes in East Africa. In addition, it was soon recognized that European domestic animals (cattle, horses, goats, etc.) were highly susceptible to this disease. Trypanosomiasis therefore placed great constraints on the development of tropical Africa by the colonial powers. Once the etiological agent and the basic life cycle of the parasite were known it became possible to initiate control programs. By 1905 the first trypanocidal drug (Atoxyl) had been developed by Ehrlich and Hata. Through a program of medical surveillance, treatment and vector control the disease was gradually brought under control, and by the mid 1970s less than 20,000 new cases per year were believed to occur. However, today human African trypanosomiasis (HAT) is recognized as a re-emerging human infection with estimates of between 350 to 500 thousand infected individuals. It must be remembered that in the absence of treatment HAT is believed to be 100% fatal. It is stated that the economic costs of the human and animal disease to the endemic area (an area approximately the size of the U.S.) is approximately 5 billion dollars per year (Coetzer, 1998).

At the present time there are 4 drugs used in the treatment of the human infection and an equally small number used for animal trypanosomiasis. The chemotherapeutic agents used for the treatment of HAT are Suramin developed in 1916, Pentamidine developed in the 1930s, Melarsoprol developed in 1949, and more recently DFMO (difluoromethylornithine) approved for use in the early 1990s. Pentamidine is used to treat patients infected with *Trypanosoma brucei gambiense*, and Suramin for the treatment of individuals infected with *Trypanosoma brucei rhodesiense*. Following treatment with either of these drugs all patients are usually given a single course of Melarsoprol injections. Individuals in the late neurological stages of the disease are treated with several series of Melarsoprol injections. DFMO is reserved for the treatment of late stage patients in which the trypanosomes are resistant to Melarsoprol. In all cases there are significant problems in the use of these drugs. Both Suramin and Pentamidine are toxic, drug resistance has been observed, and they are expensive. Suramin which is administered intravenously (IV) is given in 6 doses every 5 to 7 days. Pentamidine must be given by the intramuscular (IM) route and requires a series of 5 to 10 injections given daily or every other day. As with Suramin, Melarsoprol must be injected intravenously and 9 to 12 doses over a period of 3 to 4 weeks are required. Melarsoprol, a critical chemotherapeutic agent because of its ability to cross the blood brain barrier and to treat late secondary stage infections, is highly toxic. It is basically insoluble in water and must be dissolved in propylene glycol. It has been estimated that 5% of the treated secondary stage patients die from drug toxicity alone. Because of Melarsoprol's toxicity it can only be administered in a hospital-type setting. In addition it has been stated that up to 20% of the treated patients currently show some form of clinical resistance to Melarsoprol (Barrett, 1999). Finally, Melarsoprol is expensive ($50.00/Course of treatment).

The only other drug available for the treatment of secondary stage patients is DFMO, an ornithine decarboxylase inhibitor. This drug can also cross the blood brain barrier and is reserved for the treatment of Melarsoprol resistant cases. It can be given orally but is most effective when administered IV. A 14-day treatment regime is required. The major problem with DFMO is that for reasons that are not fully understood, *T.b.rhodesiense* shows a natural resistance to DFMO. Drug resistance to DFMO can rapidly be developed in the laboratory by the trypanosomes. It is therefore possible that further drug resistance will be observed in the field. Finally, it is also very expensive ($250/patient for the drug alone) in comparison to the other chemotherapeutic agents.

It should be noted that similar problems exist for those drugs currently used for the treatment of animal trypanosomiasis. Drug resistance is a problem, they must be injected, and they are expensive for African farmers.

Although research on the development of new trypanocidal drugs is being performed, it is extremely limited. Table 1 gives several examples of recently discovered compounds that are

Table 1. *A comparison of the IC$_{50}$ of different chemotherapeutic agents against African trypanosomes grown in vitro.*

DRUG	ORGANISM(s)/IC$_{50}$	
	STIB 900 (*T.b.r.*)	
Suramin[1]	7.49 nM	
Melarsoprol[1]	1.76 nM	
	Lab EATRO 110	KETRI 243
	(*T.b.b.*)	(*T.b.r.*)
Melarsen Oxide[2]	4.40 nM	13.00 nM
Diaminotriazine derivative SIPI-1029	2.15 nM	0.60 nM
5-deoxy-5-(hydroxy-ethylthio)adenine(HETA)[3]	0.75 µM	
	LOUTat 1 (EATRO-1886) (*T.b.r.*)	
Pentamidine HCL[4]	0.48 nM	
Pentamidine Isethionate[4]	0.67 nM	0.38 ng/ml
1,4Di(3'-amidino-phenoxy) pentane[4]	0.55 nM	
Diamidine analogues		
DB-755	0.80 nM	
DB-289	>50 µM	
Plant Extracts	STIB 900 (*T.b.r.*)	
Bussea occidentalis[1]	0.5 mg/ml	

Physalis angulata[1]	1.3 mg/ml
	ILTat 1.1 (*T.b.b.*)
MePip-Phe-hphe-vs[6]	0.10 μM
(methyl piperazine urea-	
phenyl-alanine-	
homophenylalanine-vinyl	
sulfone-phenyl	
Cbz-phe-phe-cmk[6]	3.60 μ
(Carbobenzoxy-	
phenylalanine-	
phenylalanine-chloromethyl	
ketone	

1. *Freiburghaus et al., 1996.*
2. *Bacchi et al., 1998.*
3. *Bacchi et al., 1991.*
4. *Keku et al., 1995.*
5. *Boykin, D., J.E. Hall, J.R. Seed and R. Tidwell, Unpublished data. DB-289 is the prodrug of DB-75.*
6. *Troeberg et al., 1999.*

believed to have potential as anti-trypanosomal agents. The search for new trypanocides varies from the screening of herbal medicines to the search for new inhibitors of critical trypanosome enzymes i.e., protease inhibitors, glycolytic pathway inhibitors and new inhibitors of enzymes of the polyamine pathway. However, there are few new compounds that are currently being developed for actual use in the treatment of African trypanosomiasis.

DRUG DEVELOPMENT PROTOCOLS:
Molecular Targets Approach.

In reading the current literature or attending scientific conferences or workshops, it becomes obvious that current funding agents are primarily concentrating on what is referred to as a molecular targets approach and often described as "rationale drug design". Basically this protocol involves identifying critical enzymes (or targets), i.e. enzymes involved in glycolysis, polyamine or nucleic acid synthesis in the African trypanosomes that differ from their host's enzymes. These enzymes are then biochemically characterized, purified and if possible crystallized. The purified enzymes are studied by NMR spectroscopy or X-ray crystallography and a 3-dimensional map of the enzyme's active sites and co-factors are produced. Based on the characteristics (size, shape, charge, etc.) of the active and/cofactor sites, lead compounds are designed and synthesized that the chemist believes will be capable of fitting into the site and inhibiting enzyme activity. To have potential as a drug these inhibitors or ligands must be able to distinguish between the active sites of the host and parasite enzymes. A specific parasite enzyme inhibitor is, however, not necessarily a chemotherapeutic agent. Further research is required to demonstrate that the inhibitor will cross the parasite surface and reach the active site; have low host toxicity, and not be catabolized to an inactive compound by either the parasite or host.

Molecular targeting or "rationale drug design" studies have already yielded an enormous amount of extremely valuable biochemical data. For example, the glycolytic pathway and the structure and function of individual enzymes in this pathway have been extensively studied. These studies have given us insights into the relationship between glycolysis, substrate phosphorylation, mitochondrial function and the oxidation-reduction balance of the cell. This work has led to recent investigations into the design of adenosine analog inhibitors of glyceraldehyde 3-phosphate dehydrogenase, a key enzyme in glycolysis (Aronov et al., 1999). Douglas (1994) gives several examples of the power of this approach and additional ones are abundant in today's literature (See Science, March 17, 287: 2000). A number of pharmaceutical companies are currently investing in the new molecular tools of genomics, bioinformational techniques and proteomics in their search for new drug targets. The ultimate goal is to screen compounds directly by transcript profiling and to identify novel activities even before the biochemical target is well characterized. We certainly believe that this type of research should be expanded. The genomic, structural genomics and bioinformational approach is here (and here to stay). Unquestionably, new and novel targets will be found and a much more diverse array of inhibitors will be discovered in these high output screens.

Although considerable energy and funds have been given to molecular targeting, the only rationally designed chemotherapeutic agent which is currently used for the treatment of African trypanosomiasis is DFMO. It should be noted that DFMO was actually designed for the treatment of human malignances not the African trypanosomes. Only by the more empirical screening technique was it found to inhibit the growth of the African trypanosomes. Although a valuable compound, it has, as previously noted, a number of drawbacks, the primary one being the ease with which drug resistance might be developed. It can probably safely be stated that drug resistance will rapidly develop to any drug that inhibits a single enzymatic site (or target). This would be especially true for any parasite in which high parasitemias are obtained in the host.

Whole organism screening and drug design.

In addition to the molecular targeting approach, one finds in the recent literature, the screening of plant materials for trypanocidal activity (Freiburghaus et al., 1996), as well as the attempt to develop better treatment regimes, e.g., drug combinations, etc (Van Gompel & Vervoort, 1997; Atouguia & Costa, 1999; Table 1). All of these approaches need to be vigorously pursued, however the bottom line is that there appears to be a very limited number of promising new compounds under investigation (Croft et al., 1997). Hopefully, it is obvious that new and better chemotherapeutic agents are urgently needed for the treatment of African trypanosomiasis. Therefore, we would like to suggest another approach, one that incorporates the old concept of whole organism screening with new technologies. As an example, we will illustrate this approach using work focused on aryl diamidines which may be viewed as pentamidine analogs.

As noted the diamidine, Pentamidine (Figure 1), has been successfully used for the treatment of the Gambian form of African trypanosomiasis for approximately 70 years. It is also

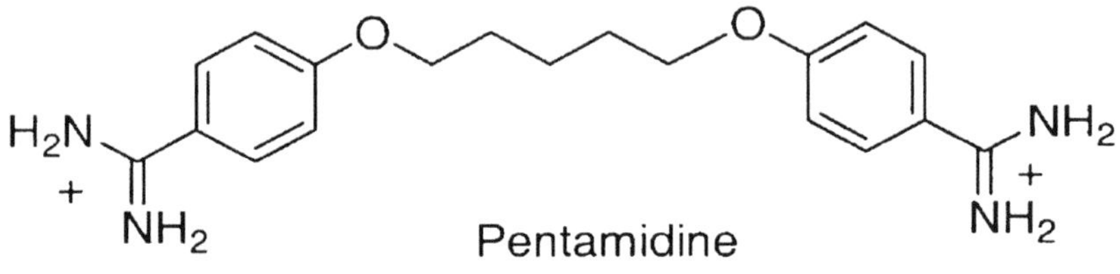

Figure 1. *Structural formula for Pentamidine.*

one of the drugs currently used for the treatment of *Pneumocystis carini*, and has been shown to inhibit the growth of a number of other protozoan parasites in vitro.

There have been extensive studies attempting to identify Pentamidine's biochemical site(s) of action. Although these studies are still on going, what is apparent is that Pentamidine has multiple target sites. It is a minor grove DNA binder (Figure 2), a DNA dependent enzyme

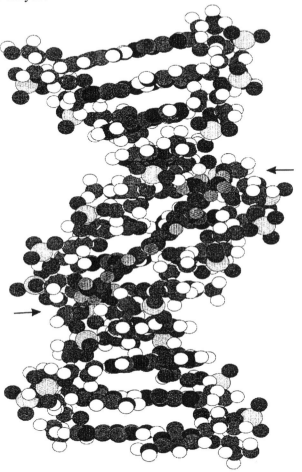

Figure 2. *Diagram of DB-75 in the DNA minor groove. The structure shows that the ligand fits tightly in the minor grove and occupies approximately four base pairs. Key hydrogen bonds are formed between the amidine hydrogens at each end of the molcule with thymine carbonyl groups of the DNA.*

inhibitor, a protease inhibitor, and inhibits several enzymes in the glycolytic pathway. Therefore resistance to Pentamidine has been slow to develop. Pentamidine has, however, a number of disadvantages and therefore the search for new and better amidine analogues has been on going for a number of years (Das and Boykin, 1977; Steck et al., 1981; Boykin et al., 1998; Francesconi et al., 1999) and continues today. Structures of some of the new classes of compounds currently being evaluated are included in Figure 3. There are libraries of untested analogues available for

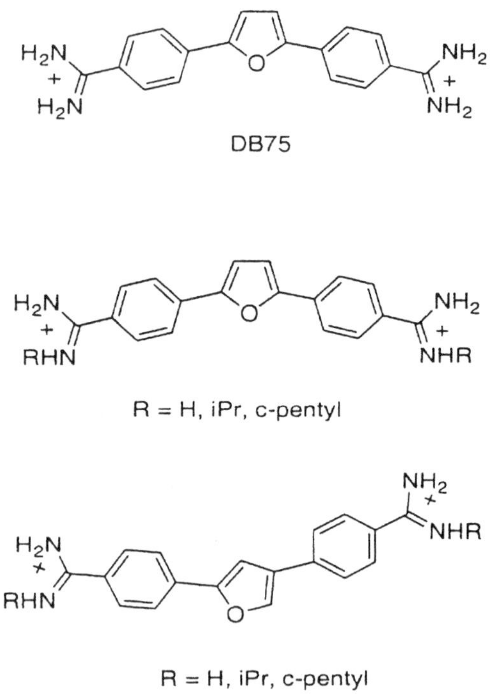

DB75

R = H, iPr, c-pentyl

R = H, iPr, c-pentyl

Figure 3. *Structural formula for DB-75 and new analogs.*

screening for trypanocidal activity. In an early preliminary study of only 16 of these analogues, 3 were found in-vitro to have significantly less toxicity to host cells, and greater trypanocidal activity than the parent compound (Keku et al., 1995). Within the last year, again in a very limited screen of 11 analogues, it has been found that one additional compound (DB-75) in this library, when administered in a single IV dose, cured mice with parasitemias of greater than 10^8/ml of blood. Mice were considered cured if trypanosomes were not detected in their blood for 20 days following treatment. In addition a pro-drug analogue (DB-289) of DB-75, also using a single oral dose cured mice with an identical acute infection (Table 2). DB-289 is referred to as a pro- drug, since the compound is totally inactive against the blood trypanosomes in vitro (Table 1) and must be enzymatically activated by the host to the parent trypanocidal compound (DB-75) (Figure 4).

Table 2. *A comparison of the in vivo activity of chemotherapeutic agents in acute mouse models of African trypanosomiases.*

DRUG	ORGANISM	DOSE (mg/kg)[1]	NUMBER OF DOSES[1]	ROUTE[1]
Diaminotriazine derivative				
SIPI-1029[2]	*T.b.b.* KETRI 243	5.0	3	ip.
SIPI-1029[2]	*T.b.b.* EATRO 110	10.0	1	ip.
SIPI-1029[2]	*T.b.b.* EATRO 110	0.5	3	ip.
HETA[4]	*T.b.b.* EATRO 110	>100.0	7	Mp[3].
Pentamidine isethionate[5]	T.b.r. LOUTat 1 (EATRO 1886)	10.0	1	ip.
1,4Di(3'-amidinophenoxy) butane[5]	T.b.r. LOUTat 1 (EATRO 1886)	5.0	1	ip.
1,5Di(3'-amidinophenoxy) pentane[5]	T.b.r. LOUTat 1 (EATRO 1886)	5.0	1	ip.
Pentamidine analogues[6]				
DB-75	T.b.b. S427	1.9	1	iv.
DB-289	T.b.b. S427	14.9	1	Oral

1   *Amount, number of days treated and route of drug administration. Animals negative for trypanosomes by tailblood examination 20 days after treatment were defined as cured.*
2   *Bacchi et al., 1998.*
3   *Mp, Implanted minipump.*
4   *Bacchi et al., 1991.*
5   *Keku et al., 1995.*
6   *Boykin, D., J.E. Hall, J.R. Seed and R. Tidwell, Unpublished data.*

The pro-drug is transported across the gut wall and then catabolized to the active compound. Furthermore, promising results with DB-75 and DB-289 have been obtained in a late stage CNS mouse model of African trypanosomiasis. The prodrug (DB-289) or the active compound (DB-75) can cross the blood-brain barrier. What we hope is obvious is that by rationally modifying the structure of a known trypanosome chemotherapeutic agent, new and better compounds can likely be obtained. The advantages of starting with a known chemotherapeutic agent are: (1) the basic biological properties of the parent compound are already known; (2) most trypanocidal drugs appear to be multiple site inhibitors and therefore the development of drug resistance should be less of a problem; (3) the basic structure and synthesis of the drug is already known and can act as a starting point for the synthetic chemist; and (4) the parent compounds have already been tested for toxicity and shown to be effective in the treatment of HAT.

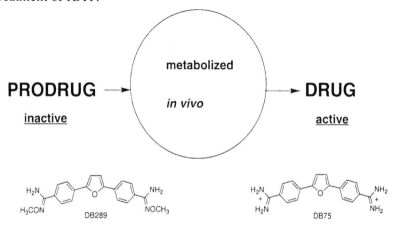

Figure 4. *Illustration of the Prodrug concept.*

Based on the proceeding discussion, we suggest that increased emphasis be placed on the synthesis and whole organism screening of rationally developed analogues of currently known trypanocidal agents. The new in vitro culture techniques would allow the screening of 100s of different compounds per week. The data obtained can be analyzed for important structural elements and new classes of lead compounds, with either reduced toxicity and/ or increased trypanocidal activity, can be developed.

During the past decade combinatorial chemistry has become a major component of virtually all drug discovery efforts. This methodology allows the rapid synthesis of hundreds and thousands of analogs and thereby, theoretically, produces the optimum compound. For example, a multiple step synthesis using robotics can be achieved in the equivalent of a 96-well plate and then the compounds can be made available for rapid screening against a variety of targets. To illustrate the potential of analog generation by combinatorial chemistry consider the synthesis of the molecule A-B-C, which involves the linking of the three molecular components A, B, and C and if three different variations each of A (A', A", etc.), B, and C are employed than a "library" of 27 compounds can be generated (3x3x3 = 27). If 10 different variations of each of the components A, B, and C are available then a "library" of 1000 compounds can be generated (10x10x10 = 1000). Thus, for moderately large molecules, which can be synthesized from a number of components (with numerous variations of the components available), libraries of tens of thousands of compounds can be made. Once a sufficient number of compounds are synthesized and screened, a limited number of promising analogues can be selected for further research into their site(s) of action, i.e., DNA binding, etc. as well as for crystallization and structure analysis. Based upon these studies it should be possible, using computer models, to predict the size and shape of the inhibitory site and then to predict the nature of the actual site, i.e., DNA binding, protease inhibitor. Ultimately it should be possible to develop novel drugs with greater ability to cross the gut wall, penetrate the blood-brain barrier, have greater inhibitory site specificity, etc. This protocol is the reverse of the current approach, but it is also rational and is predicted too more rapidly discover new chemotherapeutic agents. Currently we are screening for analogues that are active against pentamidine resistant clones of trypanosomes; analogues that are effective when administered orally; and compounds that are able to cross the blood-brain barrier.

We are not suggesting the blind screening of thousands of compounds. Rather we are proposing the rationale creation of selected libraries of different classes of analogues of the currently used chemotherapeutic agents. This approach differs from earlier screening programs in that it is now possible to rapidly and inexpensively screen many more compounds for trypanocidal activity and host cell toxicity. There are also new techniques for examining both blood-brain transport and drug pharmacokinetics. Finally, the protocol starts with a known proven chemotherapeutic agent and, using the power of modern chemistry and computer modeling, permits the rapid synthesis of many new classes of analogues. Remember, the old empirical approach was successful. It led to the series of compounds that we now have for the treatment of HAT and animal trypanosomiasis. The protocol proposed here should both rationally and more efficiently increase the rate of drug discovery. A similar defense of enlightened empirical drug design has previously been given by Hudson (1994).

We are convinced that even within our current limited library of 800 amidine analogues, that compounds superior to DB-75 and DB-289 will be found. Currently approximately 40 diamidine analogues have been screened in vitro for trypanocidal activity and host cell toxicity out of 800 available. Of these 40, 8 have shown activity that is similar to, or better than, pentamidine. It therefore could be predicted, that out of the 760 untested analogues, there could be as many as 150 additional compounds with better chemotherapeutic potential than pentamidine. However, current funding by the major organizations appears to be primarily focused on the molecular targeting approach. The second whole organism protocol described in this paper is frequently not considered basic research and therefore less likely to be funded. We argue, as did Croft in 1994, for a more balanced approach using both the whole organism based and molecular targeted protocols for drug development against HAT. The need today for new drugs is greater than ever!

PROBLEMS IN DRUG DEVELOPMENT (OR THE BOTTOM LINE).

Even if the ideal drug to treat African trypanosomiasis was discovered it would be impossible to interest most (if not all) pharmaceutical firms to do the necessary drug toxicity studies or perform the human trials. The ability of individuals (or African nations) to pay for trypanocidal drugs is insufficient for these companies to likely make a profit. The wealthy developed nations do not have a trypanosomiasis problem and therefore pharmaceutical firms can not make up lost revenues by sales of trypanocidal drugs in the developed nations. Currently we are aware of only one drug company investing in research and development of chemotherapeutic agents active against the African trypanosomes, and this investment is extremely modest. In fact, it is our impression that the World Health Organization is finding it difficult to maintain the supply of the 4 compounds currently in use (New York Times, May, 2000). In the absence of substantial subsidies to pharmaceutical firms and research institutes, we are pessimistic that any new compounds will be available for treatment of HAT in the near future. The rich get richer, the poor get poorer, and trypanosomiasis and other parasitic diseases seem to fluorish. This was the recent topic of the Global Health

Forum, a meeting of public health experts in San Francisco, U.S.A. (reported by M. Enserink in Science 287:1571, 2000). A solution to the problem that has been suggested, is to provide incentives to the drug and vaccine development industry. It was noted that President Clinton announced a 50 million-dollar contribution to the Global Alliance for Vaccines and Immunizations plus other tax credit incentives for companies. Although a noteworthy start, the funds are apparently primarily for development of vaccines for malaria, AIDS and tuberculosis. No mention was made of funds for the development of chemotherapeutic agents against trypanosomiasis, or a host of other parasitic infections.

Finally, on an equally pessimistic note, even if the ideal drug was developed and approved for use in the treatment of African trypanosomiasis, there is still a question as to whether it could be delivered to those in need. As noted elsewhere in this text, the public health infrastructure in many endemic areas has totally disintegrated. Therefore in our opinion, drug development is currently inhibited by too narrow an approach, limited research and development funds, and socioeconomic factors leading to civil unrest. Answers to these many questions may require changes in research priorities (drug development versus public health infrastructure, etc.) and many more funds than are currently available.

ACKNOWLEDGEMENTS: The authors wish to thank Mr. J. Sechelski, Drs. J.Ed Hall, and R. Tidwell for their many helpful discussions over the years as scientific colleagues.

REFERENCES:

Aronov, A.M., S. Suresh, F.S. Buckner, W.C. van Voorhis, C.L.M.J. Verlinde, F.R. Opperdoes, W.G.J. Hol and M.H. Gelb, 1999. Structure based design of submicromolar, biologically active inhibitors of trypanosomatid glyceraldehyde 3-phosphate dehydrogenase. Proceedings National Academy of Sciences **96**: 4273-4278.

Atouguia, J. and J. Costa, 1999. Therapy of Human African Trypanosomiasis: Current situation. Mem Inst Oswaldo Cruz, Rio de Janeiro **94**:221-224.

Bacchi, C.J., J.R. Sufrin, H.C. Nathan, A.J. Spiess, T. Hannan, J. Garofalo, K. Alecia, L. Katz and N. Yarlett, 1991. 5'-alkyl-substituted analogs of 5'methylthioadenosine as trypanocides. Antimicrobial Agents and Chemotherapy **35**: 1315-1320.

Bacchi, C.J., M. Vargus, D. Rattendi, B. Goldberg and W. Zhou, 1998. Anti-trypanosomal activity of a new triazine derivative, SIPI 1029, in vitro and in model infections. Antimicrobial Agents and Chemotherapy **42**: 2718-2721.

Barrett, M.P., 1999. The fall and rise of sleeping sickness. Lancet **353**:1113-1114.

Boykin, D.W., A. Kumar, G. Xiao, W.D. Wilson, B.K. Bender, D.R. McCurdy, J.E. Hall and R.R. Tidwell, 1998. 2,5-Bis-[4(N-alkylamidino)phenyl]furans as *Pneumocystis carinii* agents. Journal Medicinal Chemistry **41**: 124-129

Coetzer, T.H.T., 1998. Proteases and phosphatases as possible pathogenesis factors in African trypanosomiasis. South African Journal of Science **94**:279-289.

Croft, S.L., 1994. A rationale for anti parasite drug discovery. Parasitology Today **10**:385-392.

Croft, S.L., J.A. Urbina and R. Brun, 1997. Chemotherapy of human leishmaniasis and trypanosomiasis. Pages 245-257, In: Trypanosomiasis and Leishmaniasis, Biology and Control. Eds Hide, G., J.C. Mottram, G.H.Coombs, and P.H. Holmes, CAB International, Wallingford, U.K.

Das, B.P. and D.W. Boykin, 1977. Synthesis and antiprotozoal activity of 2,5-bis(4-guanylphenyl) furans. Journal of Medicinal Chemistry **20**:531-536.

Douglas, K.T., 1994. Rational drug design in parasitology. Parasitology Today **10**:389-392.

Francesconi, I., W.D. Wilson, F.A. Tanious, J.E. Hall, B.K. Bender, D.R. McCurdy, R.R. Tidwell and D.W. Boykin, 1999. 2,4-Diphenyl Furan Diamidines as Novel Anti-*Pneumocystis carinii* Pneumonia Agents. Journal Medicinal Chemistry **42**: 2260-2265.

Freiburghaus, F., R. Kaminsky, M.H.H. Nkunya, and R. Brun, 1996. Evaluation of African medicinal plants for their invitro trypanocidal activity. Journal of Ethnopharmacology **55**:1-11.

Hudson, A.T., 1994. The contribution of empiricism to antiparasite drug discovery. Parasitology Today **10**:387-389.

Keku, T.O., J.R. Seed and R.R.Tidwell, 1995. The invitro HL-60 cell-*Trypanosoma brucei rhodesiense* culture system: a rapid invitro drug screen. Tropical Medicine and Parasitology **46**: 258-262.

Laughton, C. A., F. Tanious, C. M. Nunn, D. W. Boykin, W. D. Wilson and S. Neidle. 1996. A Crystallographic and Spectroscopic Study of the Complex Between d(CGCGAATTCGCG)$_2$ and 2,5 Bis(4-guanylphenyl)furan, an Analogue of Berenil. Structural Origins of Enhanced DNA-Binding Affinity, Biochemistry, 35, 5655-5660.

Rosamond, J. and A. Allsop, 2000. Harnessing the power of the genome in the search for new antibiotics. Science **287**: 1973-76.

Steck, E.A., K.E. Kinnanmon, D.S. Rane and W.L. Hanson, 1981. *Leishmania donovani, Plasmodium berghei*, and *Trypanosoma rhodesiense*: antiprotozoal effects of some amidine types. Experimental Parasitology **52**: 404-413.

Troeberg, L., R.E. Morty, R.N. Pike, J.D. Lonsdale-Eccles, J.T. Palmer, J.H. McKerrow and T.H.T. Coetzer, 1999. Cysteine protease inhibitors kill cultured bloodstream forms of *Trypanosoma brucei brucei*. Experimental Parasitology **91**: 349-355.

IMMUNOBIOLOGY OF AFRICAN TRYPANOSOMIASIS: NEW PARADIGMS, NEWER QUESTIONS

John M. Mansfield, Tinsley H. Davis, Melissa E. Dubois
Department of Bacteriology, 1925 Willow Drive / FRI Building
University of Wisconsin-Madison, Madison, WI 53706 USA

ABSTRACT

The immunology of African trypanosomiasis is reviewed and reexamined in light of newer findings over the past decade. New paradigms of host resistance include information that the variant surface glycoprotein specific Ab response is not linked, alone, to overall resistance, and that Th1 cells, IFN-γ and macrophages play a powerful role in providing tissue-specific protection against trypanosomes. New questions concerning the specific role of "antigen pattern" recognition of the VSG coat by B cells, and the potentially interesting role of the trypanosome T lymphocyte triggering factor (TLTF), are addressed. Overall this chapter is aimed at providing the reader with new perspectives on the immunobiology of African sleeping sickness.

Key Words VSG-specific, Th1, INF-γ, T-lymphocyte triggering factor, host resistance, VSG mosaicism and delayed induction of immune responses.

INTRODUCTION (Figs at end of text)

Textbook lore on the immunology of African trypanosomiasis includes a nearly century-old perception that host Ab responses play the only role in controlling the disease process [see recent immunology texts; Abbas *et al.*, 2000]. Antigenic variation of the trypanosome surface coat structure, which is composed of variant surface glycoprotein (VSG) molecules, is viewed as the major mechanism employed by the parasites to evade such B cell responses and to permit survival of these organisms within the host (Vickerman and Luckins, 1969; Cross, 1975; Cross, 1990; Van der Ploeg *et al.*, 1992; Borst and Rudenko, 1994; Borst *et al*, 1996; Cross, 1996). However, the emergence of new paradigms in host resistance to trypanosomes during the past decades has been greeted with curious indifference. Largely this is due to the firm establishment of older paradigms in mainstream texts on immunology and parasitology; unfortunately, many investigators naturally turn to these resources for their base information. Thus, it is the goal of this chapter to bring the reader up to date with our current thinking on the immunology of trypanosomiasis, and to examine several new paradigms that enlarge on the historical concepts of host resistance to trypanosomes.

NEW PARADIGMS

There are clear differences in the ability of various host species, and strains within species, to display relative resistance to African trypanosomiasis (Mulligan, 1970; Levine and Mansfield, 1981. Studies over the past twenty years have revealed that the host Ab response plays only a *partial* role in such relative resistance against trypanosomes. While VSG-specific Ab clearly is responsible for the cataclysmic elimination of variant antigenic types (VATs) from the bloodstream of infected hosts (see Figure 1), it is now known that this event is not linked, functionally or genetically, to overall host resistance (De Gee and Mansfield, 1984; De Gee *et al.,* 1988; Mansfield, 1990; Mansfield, 1995; Mansfield and Olivier, 2001). The seminal studies were those in which H-2 compatible radiation chimera mice, reconstituted with reciprocal bone marrow cell transplants from relatively resistant or susceptible donors, revealed the following: that susceptible mice, which normally do not make a sufficient Ab response to VSG and do not clear VATs from the blood, were afforded by donor cells from resistant mice a functional B cell response that enabled them to clear parasitemia during infection; however, despite the ability to eliminate trypanosomes from the blood, these animals were just as susceptible as mice receiving susceptible donor bone marrow cells that failed to make protective VSG-specific B cell responses (De Gee and Mansfield, 1984). Subsequent genetic studies with crosses between Ab^+ resistant and Ab^- susceptible mouse strains showed that the F1 hybrids all were able to make VAT-specific Ab responses and control parasitemias, but all such hybrids were as susceptible as the susceptible parental strain (De Gee *et al.,* 1988). *Taken together, these types of results showed that the VSG-specific B cell response , although linked to trypanosome clearance from the blood, was not by itself functionally or genetically linked to overall host resistance.*

This information led the way to studies that first elucidated Th cell responses to VSG and other trypanosome antigens during infection (Schleifer *et al.,* 1993; Mansfield, 1994; Schopf *et al.,* 1998; Hertz *et al.,* 1998; Hertz and Mansfield, 1999). T cell responses to trypanosome antigens were not discovered previously because of several interesting characteristics of trypanosome infections. First, a nonspecific immunosuppression of T cell responses in trypanosomiasis had been recognized for many years (Mansfield and Wallace, 1974),and, although earlier studies revealed that T cell responses to trypanosome antigens could be induced in immunized animals (Finerty *et al.,* 1978; Campbell *et al.,* 1982), such responses were not readily detectable in infected animals (Mansfield and Kreier, 1972; Paulnock *et al.,* 1989). For example, not only were spleen or lymph node T cells from infected mice unable to proliferate in response to mitogens or antigen, they also failed to produce significant amounts of IL-2 or IL-4, and these events could be shown to impact on T-dependent B cell responses to a variety of antigens (Mansfield and Bagasra, 1978). This generalized immunodeficiency was shown to result in part from the presence of macrophage "suppressor cells" in lymphoid tissues (Wellhausen and Mansfield, 1979, 1980a,b; Sacks *et al.,* 1982); in fact, macrophages from infected mice had the capacity to actively suppress the proliferative responses of normal T cells to mitogens and antigens in vitro and in vivo. A breakthrough in recognizing that Th cell responses to trypanosome antigens occurred during infection came with the finding that functional compartmentalization of such responses occurred (Schleifer *et al.,* 1993). It was revealed that Th cells reactive with VSG were predominant in the peritoneal T cell population; when stimulated with VSG, these cells made a substantial IL-2 and IFN-γ cytokine response but failed to proliferate. Subsequently, it was discovered that Th cells in the peripheral lymphoid tissues also made an IFN-γ response (but little IL-2) when stimulated with VSG. Thus, it was apparent that VSG-reactive T cells were present in infected animal tissues but that they exhibited a restricted cytokine response and minimal evidence for clonal expansion (Schleifer *et al.,* 1993). Since these VSG-reactive T cells displayed a CD4$^+$ αβTCR$^+$ membrane phenotype, expressed Type 1 cytokines, were MHC II restricted and APC dependent (Schleifer *et al.,* 1993; Schopf *et al.,* 1998; Hertz *et al.,* 1998), it was clear that these represented a classical subset of Th cells that recognized VSG during infection. More recent work has begun to elucidate the submolecular targets of VSG-reactive Th cells. In preliminary studies we have shown that Th cell specificities are directed against a defined hypervariable subregion of VSGs that is not exposed when VSG homodimers are assembled into the surface coat structure (Mansfield and Olivier, 2001; Mansfield, unpublished data), fulfilling earlier predictions that VSG sequence variability in nonexposed regions of the molecule might be driven by T cell selection (Reinitz *et al.,* 1992; Blum *et al.,* 1993; Field and Boothroyd, 1996).

The extreme polarization of the Th1 cell cytokine responses seen in our experimental system is due in part to the early production of IL-12 by macrophages exposed to trypanosomes (Mansfield *et al.,* 2001). That IL-12 is not the only polarizing factor is seen from preliminary studies with IL-12 KO mice and mice exposed to Abs against IL-12; in each case, early temporal depression of the Type 1 cytokine response did not result in a compensatory Type 2 cytokine response and, after a period of 10 days or so, the Th1 cell response emerged in both groups (Mansfield *et al.*, 2001). Thus, there are profound features of infection that promote the production of Type 1 cytokines and the outgrowth of antigen-reactive Th1 cells. While reasons for the relative tissue compartmentalization of Th cell cytokine response (e.g., IL-2 and IFN-γ production by peritoneal Th cells, but mostly IFN-γ production by Th cells in the peripheral lymphoid tissues) have not been resolved, the reason for the inhibition of T cell clonal expansion has been elucidated however. Suppressor macrophages were shown to elaborate several factors that inhibited the proliferative (but not the cytokine) responses of VSG activated Th cells: nitric oxide (NO), prostaglandins and IFN-γ (Hertz and Mansfield, 1999; Sternberg and McGuigan, 1992; Schleifer and Mansfield, 1993; Darji *et al.*, 1996). Macrophages were activated to produce these suppressive factors primarily as the result of exposure to IFN-γ released by parasite antigen-activated Th cells (Hertz and Mansfield, 1999; Schleifer and Mansfield, 1993). The full impact of NO and prostaglandins on host immunity to trypanosomes has not been completely resolved, but studies with iNOS KO mice have shown that, although NO is the main "suppressor" factor that limits clonal expansion of T cells (and maybe also modulates cytokine responses to a degree) the absence of NO did not affect overall host resistance [see below; (Hertz and Mansfield, 1999)].

Trypanosome-specific Th1 cell responses may provide an essential component of host resistance; this realization emerged from studies with cytokine gene knockout mice. The central finding in our work was that mice with a resistant C57BL/6 genetic background but which lacked a functional IFN-γ gene were as susceptible as *scid* mice to trypanosome infection, even though those mice produced Abs sufficient to control parasitemia (Figure 2)(Hertz *et al.*, 1998). In contrast, BL/6 mice with the IL-4 gene knockout were as resistant as *wt* mice to infection (Hertz and Mansfield, 1999). These results underscored earlier studies demonstrating that the VAT-specific Ab response and control of parasitemia were not capable of providing resistance alone, and that the production of a single cytokine, IFN-γ, in response to infection was found to be a critical element in host resistance. The mechanism(s) associated with IFN-γ-mediated resistance are not yet clear, but seem to involve macrophage factors induced by IFN-γ activation. Several candidate factors have been proposed, such as NO and TNFα, both of which have been shown to kill trypanosomes in vitro (Vincendeau *et al.*, 1992; Mnaimneh *et al.*, 1997; Lucas *et al.*, 1993, 1994; Magez *et al.*, 1997, 1999). Recent studies suggest, however, that neither factor alone is capable of mediating resistance in vivo; results with trypanosome infected iNOS KO mice and TNFα KO mice showed that such genetic mutations on a resistant mouse genetic background did not significantly affect the course of infection (Magez *et al.,* 1999; Hertz and Mansfield, 1999; Millar *et al.,* 1999), although it is possible that the combination of NO and TNFα is required for functional resistance. Clearly, IFN-γ inducible events in macrophages must carefully be evaluated for their impact on trypanosomes during infection. Since these events occur independently of B cell mediated resistance mechanisms that are known to control trypanosomes in the vasculature, since IFN-γ activated macrophage control mechanisms are presumed to be important in regulating trypanosome numbers in the extravascular tissue spaces [but this by itself is inadequate to provide protection; (Mansfield, unpublished data)], *it appears that multiple arms of the host immune system are required to control trypanosomes and to provide relative resistance during infection.*

Currently, then, we have come to view relative resistance to African trypanosomes as being mediated by two major components of host immunity, neither one of which by itself is adequate alone to provide resistance (Figure 3). First, VSG specific Ab responses control trypanosomes present in the blood. Second, Th cell production of IFN-γ and subsequent macrophage activation events are necessary to control trypanosomes in the extravascular tissues. Animals that make weak B cell and/or T cell responses to trypanosome variant antigens invariably will demonstrate relative susceptibility; in contrast, animals making pronounced B and T cell responses (including appropriate macrophage activation events) will display relative high resistance. Further, events that impact on B, T or macrophage responses during infection can be expected to cause modulations in host resistance. The sections below explore several variations on this new paradigm of host immunity to African trypanosomes.

NEW QUESTIONS ARISE

Aside from unresolved elements of immunity to trypanosomes discussed above, there are a number of additional considerations that may modify the perspective of host resistance presented in Figure 3. Two questions that we currently are addressing in our lab are posed below as part of experiments in progress.

Does Surface Coat Antigenic Mosaicism Lead to Escape from Early B cell Recognition?

An early B cell response to the trypanosome surface coat is an important element in clearing trypanosomes from the bloodstream;

this response is quite rapid, occurring within several days of VAT outgrowth, is largely Th cell-independent, and is capable of eliminating the relevant VATs from the bloodstream (Mansfield, 1994; Mansfield *et al.*, 1981; Reinitz and Mansfield, 1988, 1990; Schopf *et al.*, 1998). This Ab response results from B cell "antigen pattern" recognition of repetitive surface epitopes that are displayed as the result of the architectural assemblage of VSG homodimers in the coat structure [(Mansfield and Olivier, 2001; Mansfield, 1994); see below). One of the major questions in trypanosomiasis is why newly emerging or even minor VATs seem to fail to trigger this protective B cell response (e.g., the kinetics of Ab elimination of each variant population largely seem to be the same for the first and subsequent waves of parasitemia, even though new VATs arise constantly throughout infection).

We previously proposed that B cells can recognize the VSG surface coat of trypanosomes by a T-independent "antigen pattern" recognition mechanism (Mansfield, 1994; Reinitz and Mansfield, 1990; Schopf *et al.*, 1998). The classical concept of pattern recognition is based on innate immune system recognition of microbial molecules by antigen-nonspecific receptors, or pattern recognition receptors, expressed on non-lymphoid cells. However, it has been shown that extensive cross-linking of B cell surface Ig receptors (BCR) by polymeric antigen structures is able to generate sufficient signals to activate B cells. A correlation between the degree of epitope repetitiveness and T-independent B cell activation has been established in viral models and general criteria have been established for the T-independent activation of B cells by polymeric antigens. Both theoretically and functionally, the VSG surface coat structure seems to fulfill these criteria as discussed below.

The orientation and packing of VSG molecules on the surface of African trypanosomes is such that approximately 10^7 identical VSG molecules are packed together on the cell membrane as homodimers. Each VSG molecule is tethered to the membrane by a GPI-anchor, with the hydrophilic N-terminal domain of the VSG oriented toward the extracellular matrix. Internal anti-parallel A and B alpha-helices provide internal molecular rigidity and also permit the close packing of VSG homodimers into the surface coat structure. Given these characteristics of the trypanosome surface coat, it has been suggested that the *architectural* assemblage of identical exposed N-terminus epitopes as part of a rigid surface structure is able to cross-link BCR and activate B cells in a T-cell-independent manner (Mansfield, 1994). This hypothesis is supported by studies demonstrating that T-independent B cell activation both *in vivo* and *in vitro* is contingent on exposure to intact, viable trypanosomes whereas, in contrast, immunization of mice or cell cultures with fixed trypanosomes or purified VSG does not elicit a T-independent VSG-specific B cell response (Mansfield, unpublished data; Reinitz and Mansfield, 1990). This suggests that the three dimensional architecture and integrity of the surface coat of viable trypanosomes is critical for B cell recognition. It also suggests that a second signal, possibly generated by viable trypanosomes, may be necessary for B cell activation.

The importance of the intact homogeneous surface coat for antigen pattern recognition by B cells has led us to examine host immune responses to trypanosomes during the process of antigenic variation, and to determine whether surface coat switching events provide an advantage to the parasites by disrupting the homogeneity of the surface coat architecture and thereby bypassing T-independent B cell activation. African trypanosomes undergo extensive antigenic variation of the VSG surface coat during the course of an infection (Vickerman and Luckins, 1969; Cross, 1975, 1990, 1996; Van der Ploeg *et al.,* 1992; Borst and Rudenko, 1994; Borst *et al.,* 1996). This occurs by sequential transcriptional activation of individual VSG genes among a large resident VSG gene family; single VSG genes are transcribed from a chromosome telomeric expression site as part of a large polycistronic transcript (Cross, 1990, 1996). During the antigenic switch process, transcription of the former VSG gene ceases and transcription of a new VSG gene is initiated from, usually, a different telomeric expression site. Both the former as well as the new VSG molecules will be displayed in the surface coat for a period of time due to stability of the previous VSG transcripts and protein coupled with synthesis, transport and diffusion of new VSG molecules into the surface coat structure. Thus, temporal double-expressor trypanosomes with a mosaic VSG surface coat periodically appear and persist for a limited time in the infected host (see Figure 4)(Borst and Rudenko, 1994; Esser and Schoenbechler, 1985; Baltz *et al.,* 1986; Timmers *et al.,* 1987; Agur *et al.,* 1989; Munoz-Jordan *et al.,* 1996; Turner, 1997). Given that these new switch variants arise in a host environment in which Ab may be present to the former coat, the questions arise whether these double-expressors enjoy some immunological advantage compared to the single-expressor trypanosomes, and whether new VSG determinants in a mosaic coat can serve as an immunological stimulus given the altered surface architecture of such a coat. We formally propose that the temporary period in which the trypanosome expresses two VSG molecular species on its surface is of interest because it may delay early B cell recognition of the mosaic coat (or even prevent clearance of dual expressor populations in which one of the surface coat VSG species had induced an existing antigen pattern specific B cell response), thus providing the parasite with a degree of anonymity (or reduced clearance) until it develops the full homogeneous coat containing only one VSG species. This question is currently being addressed in our lab using molecular techniques originally developed by the Cross laboratory that have enabled investigators to transform bloodstream form *T. brucei* with exogenous VSG gene constructs, making it possible to stably express two VSG species in the coat to mimic this transition period in the process of antigenic variation (Munoz-Jordan *et al.*, 1996). These and other inquiries into T-independent B cell recognition of the trypanosome surface will provide information concerning the parasite's ability to evade host immune responses at critical periods during the trypanosome's development of new surface coats.

Does TLTF Induced IFN-γ Provide Genetically Susceptible Hosts with a Basal Level of Resistance?

Although we have determined in our studies that IFN-γ is a Th cell derived factor that provides trypanosome infected animals with a critical component of resistance [Figures 2 and 4, above; (Hertz *et al.*, 1998)], the precise role of IFN-γ in trypanosomiasis has been debated. The key element in this debate is a molecule known as T lymphocyte triggering factor (TLTF). First described by Bakhiet, Olsson and colleagues (Bakhiet *et al.*, 1990, 1993, 1996; Olsson *et al.*, 1991, 1992, 1993) and subsequently cloned by the Donelson lab, TLTF is a 453 aa protein with potentially important biological effects (Vaidya *et al.*, 1997). TLTF was discovered when researchers noted that rodents infected with *T. b. brucei*, or lymphoid cells cultured with trypanosomes in vitro, exhibited an increase in the number of antigen non-specific IFN-γ secreting cells; depletion of $CD8^+$ T cells in animals or cultures abrogated the effect and, interestingly, also resulted in less trypanosome growth (Bakhiet *et al.*, 1990). Use of a chamber system separating lymphoid cells and trypanosomes showed that a soluble factor was responsible for induction of IFN-γ synthesis (Olsson *et al.*, 1991).

Several *Trypanosoma* species appear to express TLTF but may possess different IFN-γ stimulating abilities as measured by the relative increase of IFN-γ producing cell numbers in the presence of extracts or culture filtrates of several species, including *T. evansi, T. b. rhodesiense*, and *T. b. gambiense* (Bakhiet *et al.*, 1996). Subsequent characterization of $CD8^+$ T cell IFN-γ activation by TLTF showed that tyrosine protein kinases are necessary for activation but protein kinase C and protein kinase A specifically are not (Bakhiet *et al.*, 1993b). Interestingly, TLTF may stimulate other cells to release IFN-γ, such as rat dorsal root ganglia and this secretion apparently also is dependent on tyrosine kinase(s) (Eltayeb *et al.*, 2000). These types of studies and their experimental extension over the past decade have led investigators to posit the following hypotheses with respect to the role of TLTF: trypanosomes secrete TLTF which binds to CD8 molecules expressed on $CD8^+$ T cells, thereby inducing antigen non-specific activation and production of IFN-γ; TLTF-induced release of IFN-γ subsequently serves as a growth factor that promotes trypanosome growth (Bakhiet *et al.*, 1990, 1996; Olsson *et al.*, 1991, 1993; Vaidya *et al.*, 1997; Hamadien *et al.*, 1999). Thus a factor secreted from the parasite, TLTF, is visualized as inducing an essential trypanosome growth factor, IFN-γ, from host cells.

Experiments partially in conflict with these proposed hypotheses exist, however. First there has been no independent confirmation that IFN-γ serves as a growth factor for African trypanosomes; unpublished studies from several labs have failed to substantiate the claim that IFN-γ can in any way serve as a growth factor and no critical studies of IFN-γ uptake or binding to trypanosomes have yet been published. Furthermore, parasitemias are higher in IFN-γ knockout mice, which are highly susceptible rather than more resistant to infection with *T. b. rhodesiense* [see Figure 2; (Hertz *et al.*, 1998)]. One might reasonably expect that parasitemias would be lower in IFN-γ deprived animals if this cytokine served in any substantial manner as a growth factor. Additionally, TLTF expression is identical in trypanosomes expressing high and low virulence attributes [Figure 5; (Mansfield, unpublished data)], suggesting that there is no modulation of the gene or protein in organisms known to exhibit rapid growth characteristics and virulence attributes for mammalian hosts.

Initial studies asserted that TLTF was a secreted protein (Vaidya *et al.*, 1997), but subsequent characterizations showed that the amino acid sequence does not contain a hydrophobic region typical of membrane-transported trypanosome proteins, though it did appear to be targeted to the flagellar pocket region through an apparent conformational signal dependent upon a specific 144 amino acid domain (Hill et al., 1999). To date there is no convincing biochemical evidence that TLTF is a secreted protein, though the gene does possess unique internal targeting sequences. More recent studies on the cell biology of TLTF have suggested an alternate (or coincident) role for the protein. The gene sequence identified as TLTF is expressed in both insect and bloodstream forms of *T. b. brucei* and the protein appears to be tightly associated with the flagellar cytoskeleton (present in detergent-resistant and $Ca2^{++}$-resistant cytoskeletal fractions of trypanosome extracts) (Hill *et al.*, 2000); modification of TLTF gene expression in the insect culture form resulted in an unusual motility defect, suggesting that TLTF may be an integral part of the trypanosome cytoskeletal architecture. Surprisingly, TLTF-like genes are present in a number of divergent eukaryotes including *Drosophila* and zebra fish. Notably, the human growth arrest specific gene GAS11, closely related to TLTF (Vaidya *et al.*, 1997), is a possible tumor suppressor molecule with a submolecular region that may localize to cellular microtubules.

Thus, it is difficult to see how TLTF, a tightly bound cytoskeleton-associated molecule, would be secreted or released in biologically active levels during infection or in cell cultures containing viable trypanosomes to affect the release of IFN-γ from host cells. Yet, it is clear that trypanosome infections and trypanosome extracts *are* capable of inducing IFN-γ release from host lymphoid cells in an antigen nonspecific manner; the levels are low and occur independently of antigen specific induction of IFN-γ from host Th1 cells (Mansfield, unpublished data; Schopf *et al.*, 1998). Thus, IFN-γ secretion induced by parasite material(s) has been a repeatable phenomenon and is clearly of some interest; there is the distinct possibility that release of biologically active TLTF (or a similar molecule with closely related effects) occurs during periods of cataclysmic elimination of VATs by host Ab and Th1/macrophage cell responses throughout infection, rather than by an active secretory pathway, to induce IFN-γ. Given that trypanosomes/trypanosome extracts are capable of inducing nonspecific IFN-γ release from host cells, and that IFN-γ is such a critical factor in host protection(probably important in regulating parasite numbers in the extravascular tissues; see Figures 2 and 3), we would like to reexamine TLTF in a different context. We formally propose here that TLTF (and/or related factors) were evolutionarily retained by trypanosomes in order to induce a basal level of IFN-γ-dependent host protection. This basal protection would take advantage of the expression of target CD8 molecules on a variety of host cell types (including CD8+ T cells) in all animals, would be independent of genetically-based host *immune* resistance factors (e.g., genetically determined levels B cell and Th1 cell/macrophage responses), but would be sufficient to prevent rapid host death following infection. Given that in nature the trypanosome has no control over what type of host (mammalian species, strain; relatively resistant, susceptible) the tsetse fly bites and to which it transmits the infection, it would seem evolutionarily expensive for trypanosomes not to have developed a mechanism to prevent host death for a period of time sufficient to permit trypanosome growth and subsequent transmission of the parasite. We propose that TLTF serves this role by inducing low levels of IFN-γ that control an early potentially explosive spread of trypanosomes throughout host tissues, regardless of the genetically-based resistance status of the host infected. Such a control over massive parasite expansion and, in susceptible animals or individuals, a delay in host death, would permit the host to survive for a period of time so that the possibility of trypanosomes being taken up in a blood meal and transmitted to new hosts could occur.

Regardless of one's view of TLTF, the hard work of providing functional and genetic linkages between TLTF expression and biological effects on the host remain. In our lab we currently are knocking out the TLTF locus (as well as overexpressing the gene) in bloodstream *Trypanosoma brucei rhodesiense* in order to examine both the cell biological and immunological aspects of TLTF. We believe this type of approach will go a long way towards addressing the biological relevance of this interesting molecule.

Summary

We hope that we have provided the reader with a new view of the immunobiology of trypanosomiasis, and have modified the paradigm of host resistance that persists in many textbooks. The ideas presented in this chapter are meant not to become the new dogma in this field, but rather to stimulate thought about the complex and highly interesting interactions that occur between the African trypanosomes and their mammalian hosts.

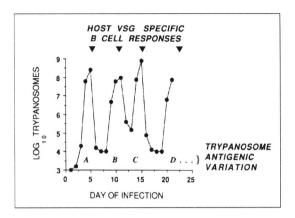

Figure 1. *Parasitemia in a mouse infected with Trypanosoma brucei rhodesiense LouTat 1. The fluctuations of trypanosome numbers in the blood are caused by Ab responses to the VSG surface coat, leading to immune elimination of organisms from the vascular tissue compartment and by the ability of the parasites to undergo antigenic variation of the VSG, permitting the organisms to repopulate tissues.*

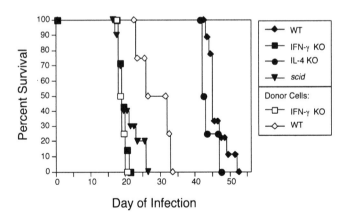

Figure 2. *IFN-γ is a critically important resistance factor in African trypanosomiasis. Infection of IFN-γ gene knockout mice resulted in survival times that were no different than scid mice, despite the fact that the knockout mice produced VSG specific Ab responses and controlled parasites in the blood (from Hertz et al., 1998. J. Immunology).*

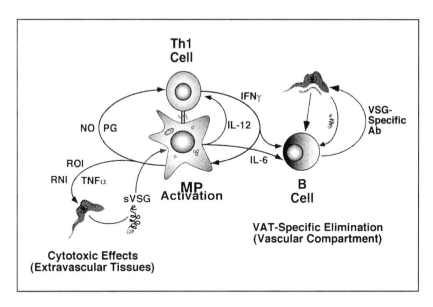

Figure 3. *Relative resistance to African trypanosomiasis is dependent upon both a B cell response sufficient to control trypanosomes in the vascular compartment and a Th1 cell/ IFN-γ/macrophage activation response to control trypanosomes in the extravascular tissues.*

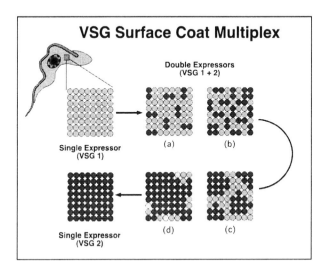

Figure 4. *Molecular mosaicism of the VSG surface coat occurs during the period when trypanosomes of one VAT switch VSG gene expression and begin to express the new VSG gene and surface coat glycoproteins of a different VAT. This temporal mosaic coat is predicted to give switch variants an immunological advantage.*

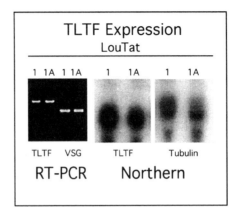

Figure 5. *Expression of TLTF in Trypanosoma brucei rhodesiense LouTat 1 and LouTat 1A, clonally related organisms that differ markedly in their virulence for mammalian hosts (Inverso et al., 1988; Mansfield, unpublished data). TLTF protein expression also appears to be similar in these organisms (Hill, personal communication).*

REFERENCES

Abbas, A. K., A. H. Lichtman, and J. S. Pober. 2000. Cellular and Molecular Immunology, 4th Edition:359.

Agur, Z., D. Abiri, and L. H. Van der Ploeg. 1989. Ordered appearance of antigenic variants of African trypanosomes explained in a mathematical model based on a stochastic switch process and immune-selection against putative switch intermediates. Proceedings of the National Academy of Sciences of the United States of America **86**:9626.

Baltz, T., C. Giroud, D. Baltz, C. Roth, A. Raibaud, and H. Eisen. 1986. Stable expression of two variable surface glycoproteins by cloned Trypanosoma equiperdum. Nature **319**:602.

Munoz-Jordan, J. L., K. P. Davies, and G. A. Cross. 1996. Stable expression of mosaic coats of variant surface glycoproteins in Trypanosoma brucei. Science **272**:1795.

Bakhiet, M., P. Buscher, R. A. Harris, K. Kristensson, H. Wigzell, and T. Olsson. 1996. Different Trypanozoan Species Possess Cd8 Dependent Lymphocyte Triggering Factor-Like Activity. Immunology Letters **50**:71.

---------------, E. Mix, K. Kristensson, H. Wigzell, and T. Olsson. 1993. T cell activation by a Trypanosoma brucei brucei-derived lymphocyte triggering factor is dependent on tyrosine protein kinases but not on protein kinase C and A. European Journal of Immunology **23**:1535.

---------------, T. Olsson, C. Edlund, B. Hojeberg, K. Holmberg, J. Lorentzen, and K. Kristensson. 1993. A Trypanosoma brucei brucei-derived factor that triggers CD8+ lymphocytes to interferon-gamma secretion: purification, characterization and protective effects in vivo by treatment with a monoclonal antibody against the factor. Scand J Immunol **37**:165.

---------------, T. Olsson, J. Mhlanga, P. Buscher, N. Lycke, P. H. Vandermeide, and K. Kristensson. 1996. Human and Rodent Interferon-Gamma As a Growth Factor For Trypanosoma Brucei. European Journal of Immunology **26**:1359.

--------------, T. Olsson, P. Van der Meide, and K. Kristensson. 1990. Depletion of CD8+ T cells suppresses growth of Trypanosoma brucei brucei and interferon-gamma production in infected rats. Clin Exp Immunol **81**:195.

Blum, J. L., J. A. Down, A. M. Gurnett, M. Carrington, M. J. Turner, and D. C. Wiley. 1993. A structural motif in the variant surface glycoproteins of Trypanosoma brucei. Nature **362**:603.

Borst, P., and G. Rudenko. 1994. Antigenic variation in African trypanosomes. Science **264**:1872.

--------, G. Rudenko, M. C. Taylor, P. A. Blundell, F. Vanleeuwen, W. Bitter, M. Cross, and R. McCulloch. 1996. Antigenic Variation In Trypanosomes. Archives of Medical Research **27**:379.

Campbell, G. H., K. M. Esser, and S. M. Phillips. 1982. Parasite- (antigen) specific stimulation of B and T cells in African trypanosomiasis. J Immunol **129**:1272.

Cross, G. A. 1975. Identification, purification and properties of clone-specific glycoprotein antigens constituting the surface coat of Trypanosoma brucei. Parasitology **71**:393.

----------. 1990. Cellular and genetic aspects of antigenic variation in trypanosomes. Annu Rev Immunol **8**:83.

----------. 1996. Antigenic variation in trypanosomes: secrets surface slowly. Bioessays **18**:283.

Darji, A., A. Beschin, M. Sileghem, H. Heremans, L. Brys, and P. Debaetselier. 1996. In Vitro Simulation Of Immunosuppression Caused By Trypanosoma Brucei - Active Involvement Of Gamma Interferon and Tumor Necrosis Factor In the Pathway Of Suppression. Infection & Immunity **64**:1937.

De Gee, A. L., and J. M. Mansfield. 1984. Genetics of resistance to the African trypanosomes. IV. Resistance of radiation chimeras to Trypanosoma rhodesiense infection. Cell Immunol **87**:85.

------------, R. F. Levine, and J. M. Mansfield. 1988. Genetics of resistance to the African trypanosomes. VI. Heredity of resistance and variable surface glycoprotein-specific immune responses. J Immunol **140**:283.

Eltayeb, R., A. Sharafeldin, R. Jaster, T. Bittorf, E. Mix, and M. Bakhiet. 2000. Trypanosoma brucei brucei induces interferon-gamma expression in rat dorsal root ganglia cells via a tyrosine kinase-dependent pathway. Journal of Infectious Diseases **181**:400.

Esser, K. M., and M. J. Schoenbechler. 1985. Expression of two variant surface glycoproteins on individual African trypanosomes during antigen switching. Science **229**:190.

Field, M. C., and J. C. Boothroyd. 1996. Sequence divergence in a family of variant surface glycoprotein genes from trypanosomes: coding region hypervariability and downstream recombinogenic repeats. Journal of Molecular Evolution **42**:500.

Finerty, J. F., E. P. Krehl, and R. L. McKelvin. 1978. Delayed-type hypersensitivity in mice immunized with Trypanosoma rhodesiense antigens. Infection & Immunity **20**:464.

Hamadien, M., N. Lycke, and M. Bakhiet. 1999. Induction of the trypanosome lymphocyte-triggering factor (TLTF) and neutralizing antibodies to the TLTF in experimental african trypanosomiasis. Immunology **96**:606.

Hertz, C. J., H. Filutowicz, and J. M. Mansfield. 1998. Resistance to the African trypanosomes is IFN-gamma dependent. Journal of Immunology **161**:6775.

--------------, and J. M. Mansfield. 1999. IFN-gamma-dependent nitric oxide production is not linked to resistance in experimental African trypanosomiasis. Cellular Immunology **192**:24.

Hill, K. L., N. R. Hutchings, P. M. Grandgenett, and J. E. Donelson. 2000. T Lymphocyte-triggering factor of African trypanosomes is associated with the flagellar fraction of the cytoskeleton and represents a new family of proteins that are present in several divergent eukaryotes. Journal of Biological Chemistry **275**:39369.

------------, N. R. Hutchings, D. G. Russell, and J. E. Donelson. 1999. A novel protein targeting domain directs proteins to the anterior cytoplasmic face of the flagellar pocket in African trypanosomes. Journal of Cell Science **112**:3091.

Inverso, J. A., A. L. De Gee, and J. M. Mansfield. 1988. Genetics of resistance to the African trypanosomes. VII. Trypanosome virulence is not linked to variable surface glycoprotein expression. J Immunol **140**:289.

Levine, R. F., and J. M. Mansfield. 1981. Genetics of resistance to African trypanosomes: role of the H-2 locus in determining resistance to infection with Trypanosoma rhodesiense. Infect Immun **34**:513.

Lucas, R., S. Magez, B. Songa, A. Darji, R. Hamers, and P. de-Baetselier. 1993. A role for TNF during African trypanosomiasis: involvement in parasite control, immunosuppression and pathology. Res Immunol **144**:370.

------------, S. Magez, R. De-Leys, L. Fransen, J. P. Scheerlinck, M. Rampelberg, E. Sablon, and P. De-Baetselier. 1994. Mapping the lectin-like activity of tumor necrosis factor. Science **263**:814.

Magez, S., M. Geuskens, A. Beschin, H. Delfavero, H. Verschueren, R. Lucas, E. Pays, and P. Debaetselier. 1997. Specific Uptake Of Tumor Necrosis Factor-Alpha Is Involved In Growth Control Of Trypanosoma Brucei. Journal of Cell Biology **137**:715.

------------, M. Radwanska, A. Beschin, K. Sekikawa, and P. De Baetselier. 1999. Tumor necrosis factor alpha is a key mediator in the regulation of experimental Trypanosoma brucei infections. Infection & Immunity **67**:3128.

Mansfield, J. M. 1990. Immunology of African trypanosomiasis. In *Modern Parasite Biology: Cellular, Immunological and Molecular Aspects*. D. J. Wyler, ed. W. H. Freeman and Co., New York, p. 222.

------------------. 1994. T-cell responses to the trypanosome variant surface glycoprotein: A new paradigm? Parasitol Today **10**:267.

------------------. 1995. Immunobiology of African trypanosomiasis: A revisionist view. In *Molecular Approaches to Parasitology*. J. C. Boothroyd, and R. Komuniecki, eds. Wiley-Liss, New York, p. 477.

------------------, and O. Bagasra. 1978. Lymphocyte function in experimental African trypanosomiasis. I. B cell responses to helper T cell-independent and - dependent antigens. J Immunol **120**:759.

------------------, and J. P. Kreier. 1972. Tests for antibody- and cell-mediated hypersensitivity to trypanosome antigens in rabbits infected with Trypanosoma congolense. Infect Immun **6**:62.

------------------, R. F. Levine, W. L. Dempsey, S. R. Wellhausen, and C. T. Hansen. 1981. Lymphocyte function in experimental African trypanosomiasis. IV. Immunosuppression and suppressor cells in the athymic nu/nu mouse. Cell Immunol **63**:210.

------------------, and M. Olivier. 2001. Immune evasion by parasites. In *Infection and Immunity*. A. Sher, and S. Kaufmann, eds. ASM Press.

------------------, D. M. Paulnock, C. J. Hertz, H. Filutowicz, L. R. Schopf, and J. Sypeck. 2001. IFNg-independent IL-12 production during trypanosome infection directs the outgrowth of highly polarized Th1 cell responses. Journal of Immunology Submitted.

------------------, and J. H. Wallace. 1974. Suppression of cell-mediated immunity in experimental African trypanosomiasis. Infect Immun **10**:335.

Millar, A. E., J. Sternberg, C. McSharry, X. Q. Wei, F. Y. Liew, and C. M. Turner. 1999. T-Cell responses during Trypanosoma brucei infections in mice deficient in inducible nitric oxide synthase. Infection & Immunity **67**:3334.

Mnaimneh, S., M. Geffard, B. Veyret, and P. Vincendeau. 1997. Albumin Nitrosylated By Activated Macrophages Possesses Antiparasitic Effects Neutralized By Anti-No-Acetylated-Cysteine Antibodies. Journal of Immunology **158**:308.

Mulligan, H. W. 1970. *The African trypanosomiases,* New York.

Olsson, T., M. Bakhiet, C. Edlund, B. Hojeberg, P. H. Van der Meide, and K. Kristensson. 1991. Bidirectional activating signals between Trypanosoma brucei and CD8+ T cells: a trypanosome-released factor triggers interferon-gamma production that stimulates parasite growth. Eur J Immunol 21:2447.

------------, M. Bakhiet, B. Hojeberg, A. Ljungdahl, C. Edlund, G. Andersson, H. P. Ekre, W. P. Fung-Leung, T. Mak, H. Wigzell, U. Fiszer, and K. Kristensson. 1993. CD8 is critically involved in lymphocyte activation by a T. brucei brucei-released molecule. Cell 72:715.

------------, M. Bakhiet, and K. Kristensson. 1992. Interactions between Trypanosoma brucei and CD8+ T cells. Parasitol Today 8:237.

Paulnock, D. M., C. Smith, and J. M. Mansfield. 1989. Antigen presenting cell function in African trypanosomiasis. *Alan R Liss, Inc ,New York 0:135.*

Reinitz, D. M., B. D. Aizenstein, and J. M. Mansfield. 1992. Variable and conserved structural elements of trypanosome variant surface glycoproteins. Mol Biochem Parasitol 51:119.

----------------, and J. M. Mansfield. 1988. Independent regulation of B cell responses to surface and subsurface epitopes of African trypanosome variable surface glycoproteins. J Immunol 141:620.

----------------, and J. M. Mansfield. 1990. T-cell-independent and T-cell-dependent B-cell responses to exposed variant surface glycoprotein epitopes in trypanosome-infected mice. Infect Immun 58:2337.

Sternberg, J., and F. McGuigan. 1992. Nitric oxide mediates suppression of T cell responses in murine Trypanosoma brucei infection. Eur J Immunol 22:2741.

Sacks, D. L., G. Bancroft, W. H. Evans, and B. A. Askonas. 1982. Incubation of trypanosome-derived mitogenic and immunosuppressive products with peritoneal macrophages allows recovery of biological activities from soluble parasite fractions. Infect Immun 36:160.

Schleifer, K. W., H. Filutowicz, L. R. Schopf, and J. M. Mansfield. 1993. Characterization of T helper cell responses to the trypanosome variant surface glycoprotein. J Immunol 150:2910.

-------------------, and J. M. Mansfield. 1993. Suppressor macrophages in African trypanosomiasis inhibit T cell proliferative responses by nitric oxide and prostaglandins. J Immunol 151:5492.

Schopf, L. R., H. Filutowicz, X. J. Bi, and J. M. Mansfield. 1998. Interleukin-4-dependent immunoglobulin G1 isotype switch in the presence of a polarized antigen-specific Th1-cell response to the trypanosome variant surface glycoprotein. Infection & Immunity 66:451.

Timmers, H. T., T. de Lange, J. M. Kooter, and P. Borst. 1987. Coincident multiple activations of the same surface antigen gene in Trypanosoma brucei. Journal of Molecular Biology 194:81.

Turner, C. M. R. 1997. Trypanosomes With Multicoloured Coats. Parasitology Today 13:247.

Vaidya, T., M. Bakhiet, K. L. Hill, T. Olsson, K. Kristensson, and J. E. Donelson. 1997. The Gene For a T Lymphocyte Triggering Factor From African Trypanosomes. Journal of Experimental Medicine 186:433.

Van der Ploeg, L. H., K. Gottesdiener, and M. G. Lee. 1992. Antigenic variation in African trypanosomes. Trends Genet 8:452.

Vickerman, K., and A. G. Luckins. 1969. Localization of variable antigens in the surface coat of Trypanosoma brucei using ferritin conjugated antibody. Nature 224:1125.

Vincendeau, P., S. Daulouede, B. Veyret, M. L. Darde, B. Bouteille, and J. L. Lemesre. 1992. Nitric oxide-mediated cytostatic activity on Trypanosoma brucei gambiense and Trypanosoma brucei brucei. Exp Parasitol 75:353.

Wellhausen, S. R., and J. M. Mansfield. 1979. Lymphocyte function in experimental African trypanosomiasis. II. Splenic suppressor cell activity. J Immunol **122**:818.

----------------------, and J. M. Mansfield. 1980. Characteristics of the splenic suppressor cell--target cell interaction in experimental African trypanosomiasis. Cell Immunol **54**:414.

----------------------, and J. M. Mansfield. 1980. Lymphocyte function in experimental African trypanosomiasis. III. Loss of lymph node cell responsiveness. J Immunol **124**:1183.

IDENTIFYING THE MECHANISMS OF TRYPANOTOLERANCE IN CATTLE.

J. Naessens, D.J. Grab*, M. Sileghem**

International Livestock Research Institute, P.O. Box 30709, Nairobi, Kenya
* Johns Hopkins University School of Medicine, Department of Pediatrics, Baltimore, USA
** I.W.T., Bischoffsheimlaan 25, 1000 Brussels, Belgium

ABSTRACT

Some West African cattle breeds such as N'Dama have survived in tsetse-infected areas for thousands of years and are productive under trypanosome challenge, a trait known as trypanotolerance. Elucidation of the mechanisms of trypanotolerance could lead to new options for disease control. This review describes responses in trypanotolerant and trypanosusceptible breeds of cattle, and compares them with responses in mouse models. It focuses on the roles of haemopoietic tissue, T lymphocytes and antibodies in resistance to trpanosomiasis.

Key words N'Dama, trypanotolerance, *T. congolense, T. vivax,* parasitemia, anemia, chimera, T cell depletion.

TRYPANOTOLERANCE IN CATTLE

The common view held amongst scientists is that present cattle populations on the African continent are derived from three main phases of introduction (Epstein H. and Mason I.L., 1983). The first two were *Bos taurine* cattle that entered the continent from middle Eurasia through Egypt. The humpless Hamitic longhorns were introduced in the Nile delta around 6000 BC, while the humpless shorthorns supposedly arrived around 2500-2750 BC. The longhorns, and later the shorthorns, dispersed through the continent, one route of migration going west along the mediterranean coast, and then South to the tsetse belt, where trypanosome challenge probably prevented further migration. A third wave of migration into Africa of humped zebu *Bos indicus* occurred mainly around 699 AD with the Arab invasion of Africa. Today, more than 150 breeds of cattle of taurine, indicine or intermediate types (Sanga cattle) inhabit the subsaharan continent. Although *B. indicus* is a late arrival in the continent, indicus characteristics are now widespread (Hanotte et al. 2000) and it is speculated that taurine breeds have only been kept because of specific disease resistance and habitat adaptation.

Despite the wide distribution of zebu, taurine breeds predominate in the tsetse belt most likely because they possess some degree of natural resistance to trypanosomiasis. Early studies suggested that longhorn N'Dama, Baoulé, and the shorthorn Muturu were more productive under natural challenge than zebu (Murray et al. 1982). There is little doubt that the long contact of the *Bos taurine* breeds in West Africa with the parasite has been the drive for selection of more resistant traits. Trypanotolerance is not limited to cattle, and has also been described in dwarf races of sheep and goats from West Africa (Toure et al. 1983), in red masai sheep and in East African goats (Griffin and Allonby, 1979; Mutayoba et al. 1989).

Trypanotolerant N'Dama, brought to East-Africa as frozen embryos, and matched trypanosusceptible Boran cattle were directly compared under the same conditions to *T. congolense*-infection. Both breeds were equally susceptible to the establishment of infection, but the N'Dama showed superior resistance: they had a better ability to limit parasitemia, anemia and the alterations in leukocyte counts (Ellis et al. 1987; Paling, Moloo, Scott, McOdimba et al. 1991). The breeds were then compared during four sequential infections with four unrelated *T. congolense* clones (Paling, Moloo, Scott, Gettingby et al. 1991). In contrast to infected Boran, the N'Damas gained weight, spontaneously recovered without treatment, showed an ability to control the intensity and prevalence of parasitemia and an ability to resist anemia. Previous exposure with heterologous trypanosomes did not affect the course of infection, although it did reduce the severity of anemia in N'Dama, but not Boran (Williams et al. 1991).

Control of development of anemia, but not parasitemia, has a major effect on overall productivity (Trail, d'Ieteren, Feron et al. 1991). Growth and average PCV showed a degree of heritability (Trail, d'Ieteren, Maille et al. 1991).

IMMUNOPATHOLOGY AND IMMUNE RESPONSES
Non-specific antibody response

Much of what we know about trypanosome and immune system interactions has been deduced from experiments in mice. However, mice are not natural hosts of the trypanosome species that cause disease in people and domestic animals and, therefore, may respond differently to humans and cattle. In addition, because of their small size, mice rapidly develop very high levels of parasitemia possibly obviating trypanosome control mechanisms that are effective only at low parasite density in larger host species thereby triggering levels of immunopathology that do not develop in these species.

In the light of these comments, it is perhaps not surprising that studies of immune responses to trypanosomes in cattle and mice reveal some differences. For example, typanosome-induced B cell activation is less pronounced in cattle than in mice. Furthermore, it also varies among cattle breeds with the lowest responses being noted in trypanotolerant N'Dama. (Williams et al. 1996; Buza et al. 1997). With rare exceptions (Assoku and Gardiner, 1989) auto-antibody activity has not been observed in trypanosome-infected cattle, but is common in mice (Sileghem, Darji et al., 1994). Nevertheless, infected cattle do generate some antibodies that are not specific for trypanosome antigens. In the case of *T. congolense* infected cattle, these may be part of the pool of natural antibodies because they (1) belong exclusively to the IgM class, (2) appear when the number of CD5[+] B cells increases (Naessens and Williams, 1992), (3) are preferentially secreted by CD5[+] B lymphocytes (Buza et al., 1997), (4) are polyreactive, and (5) are present in low titres before infection (Buza and Naessens, 1999). Natural antibodies are produced in the absence of antigenic stimulation and may have a protective role during infection (Ochsenstein and Zinkernagel, 2000).

Variant antigen-specific antibody responses

Protective immunity in cattle can be induced by the generation of antibodies that react with surface epitopes on the parasite-attached variable surface glycoprotein (VSG) (Luckins, 1976; Musoke et al. 1981). The antibodies are variant antigenic type (VAT) specific and their epitopes are dependent on an intact VSG conformation and on the integrity of the trypanosome surface. VAT-specific antibodies mediate complement-mediated lysis (Crowe et al. 1984), agglutination (Russo et al. 1994), immobilization (Wei et al. 1990) and increased macrophage uptake (Ngaira et al. 1983) of trypanosomes. There is little if any difference in trypanodestructive VAT-specific antibody responses of trypanosome-infected trypanotolerant and susceptible cattle (Murray et al. 1982; Pinder et al. 1987; Williams et al. 1996). However, the N'Damas do have higher IgG1 responses to cryptic VSG epitopes. The superior capacity of trypanotolerant cattle to control trypanosomiasis is unlikely to be due to a faster and higher antibody response to VSG surface epitopes. Whether or not other functional attributes of antibodies, such as affinity and avidity, influence resistance phenotype and differentiate susceptibility from trypanotolerance is yet to be determined.

Non-variant antigen-specific antibody responses

In contrast to VAT-specific responses, antibody titres to non-variant antigens differ between tolerant and susceptible cattle. Antibodies to a 33 kD trypanosome antigen identified as a cysteine protease (Authié et al. 1992; Mbawa et al. 1992) called congopain, appear to be associated with trypanotolerance (Authié, Duvallet et al. 1993). Although antibodies to trypanosomal proteases and other non-variant antigens may not be involved in killing or clearance of trypanosomes, they may neutralize the pathogenic or toxic effects of their target antigens and, therefore, these antigens may be useful candidates for an "anti-disease" vaccine. "Anti-disease" antibodies, naturally acquired after repeated infections, could be responsible for the reduction of pathological effects, by providing better control of anemia after repeated heterologous infections (Paling, Moloo, Scott, Gettingby et al. 1991). Experiments to determine whether immunization to congopain induces a degree of resistance are ongoing (Authié, personal comunication).

Complement

Trypanosome infection in cattle is accompanied by severe depression of haemolytic complement activity (Nielsen et al. 1978) that is less profound in sera from trypanotolerant cattle than in sera from susceptible ones (Authié and Pobel, 1990). It is not clear whether the decrease in complement activity can be accounted for by the formation of immune complexes (Tabel et al. 1980). Because of the important role of complement, its depletion in the serum could cause further defects in the development of the immune response. These include a possible defect in the IgM to IgG switch in susceptible animals (Authié, Muteti et al. 1993; Taylor, Lutje, Kennedy et al. 1996), and an increased susceptibility of the animals to secondary infection.

Anemia.

The most significant feature of bovine trypanosome infections is the development of a profound and progressive anemia (reviewed in Murray & Dexter, 1988). Red cell breakdown begins with the first appearance of parasites in the blood and this process progresses rapidly during the next 2 to 4 weeks. Red cell destruction appears to be due to a massive erythrophagocytosis in spleen and liver. Thereafter, blood packed cell volume (PCV) values recover in trypanotolerant animals, whereas they typically remain low in susceptible animals. Unless cured by chemotherapy, the susceptible animals will die. The recovery phase in trypanotolerant cattle is variable in time of duration and coincides with a lower parasitemia. In response to the anemic state, erythropoietic activity increases, however the increase in activity fails to reach a level expected for the degree of anemia encountered (Dargie et al. 1979a: Andrianarivo et al. 1995).

T cell responses and immune suppression

Trypanosome infection causes suppression of the host immune system that is profound in mice. This suppression involves macrophages which both inhibit secretion of IL-2 and the expression of the IL-2 receptor on T cells and loss of their proliferative capacity (reviewed in Sileghem, Darji et al. 1994). Interestingly, IFN-γ production remains unaffected. Two pathways that induce suppression were identified. The first was found in lymph nodes and spleen and was mediated by prostaglandins. A second pathway was found predominantly in lymph nodes, appeared later in infection, and was mediated by an as yet unidentified mechanism. In addition to prostaglandins, other molecules have been implicated in the effector phase of suppression, such as nitric oxide (NO) (Sternberg and McGuigan, 1992). Studies on the causal role of NO revealed that infection elicits both nitric-oxide dependent and NO-independent pathways, again suggesting the existence of two pathways (Beschin et al. 1998). The relationship of these to the recently described classical and alternative pathways of macrophage activation (Goerdt and Orfanos, 1999) is not resolved.

Trypanosome-infected cattle appear to be less immunosuppressed than mice. Similar to the situation in mice, suppression in cattle was observed at the level of IL-2 production and IL-2R expression, but not at the level of IFN-γ production. Again similar to mice, suppressive cells were macrophages. In contrast to mice, no evidence was found for a prostaglandin-dependent pathway, and the prostaglandin-independent pathway in cattle was not restricted to the IL-2R and also acted at the level of IL-2 production (Sileghem & Flynn, 1992). Furthermore, unlike in mice, NO production is downregulated in infected cattle (Taylor et al. 1998) and inhibitors of NO production have no impact on the proliferation of cattle lymph node cells *in vitro*. In addition, immunosuppression in cattle is transient and restricted to the lymph nodes. Its impact on survival of the infected cattle is most likely marginal. Finally, no difference in the degree of induced immunosuppression was observed between tolerant and resistant cattle (Flynn and Sileghem, 1993). It would appear that the suppression is not a distinguishing feature of trypanosusceptibility in bovids.

Macrophage activation and co-stimulatory cytokines
        Several functions of macrophages have been shown to be
activated during African trypanosome infections in mice, including
phagocytosis, receptor-mediated phagocytosis, adherence,
pinocytosis, oxidative burst, prostaglandin release and IL-1 secretion
(Sileghem, Darji et al. 1994). In contrast, *T. congolense* infections in
cattle do not appear to activate macrophages to a high degree. The
secretion of costimulatory cytokines by spleen and blood monocytes,
measured in a thymocyte proliferation assay, was transiently
increased in both trypanotolerant N'dama and the more susceptible
Boran (Sileghem et al. 1993). IL-1 and TNF-α mRNA did not
increase throughout infection (Mertens et al. 1999). A small, but
transitory increase in the *ex-vivo* production of TNF-α by monocytes,
peaking around 2-3 weeks, was observed after an infection with *T.
vivax* in Boran cattle (Sileghem, Flynn et al. 1994). However, no such
increase was observed in Boran cattle infected with *T. congolense*.
The secretion of NO by bovine macrophages decreased during
infection (Taylor, Lutje and Mertens, 1996). As both anti-TNF-α
(Darji et al. 1996) and anti-IFN-γ (Beschin et al. 1998) can enhance
the release of NO in infected mice, it is possible that a lack of TNF-α
in infected cattle is one of the reasons for the down-regulation of NO
production.
        In general, the marked activation of macrophages observed in
infected mice does not appear to be observed in cattle. Since the
surface coat protein of the parasite has been shown to be a
macrophage-activating molecule (Magez et al. 1998; Sileghem et al.
2001), the higher parasite load in mice might be responsible for the
overwhelming activation in mice.

Other cytokines
        T. congolense-infected cattle have elevated levels of mRNA
encoding IL-10 in cells from blood, lymph node and spleen,
suggesting that IL-10 might be related to the apparent failure of
bovine monocytes to produce inflammatory molecules (Taylor et al.
1998). This cytokine has been associated with Th2-type responses
and is known to suppress expression of other cytokines and regulatory
molecules, such as NO and TNF-α (Taylor et al. 1998). The source of
the increased IL-10 message was not the macrophage population
(Mertens et al. 1999), but could be the CD5[+] B-cell population, which
expands during an infection (Naessens & Williams, 1992) and is
kinetically associated with the increase in IL-10 (Taylor et al. 1998).
The increase in IL-10 is found in both tolerant N'Dama and

susceptible Boran cattle and hence may not be associated with trypanotolerance or susceptibility.

The association of expression of a number of cytokines (IL-1α, IL-2, IL-4, IL-5, IL-6, IL-12 p40, TNF-α, CD40L, TGF-β) with disease susceptibility was investigated after infection (Mertens et al. 1999). The expression of IL-4 was upregulated in N'Dama at day 32 p.i., but not in Boran, suggesting that IL-4 may have a protective role.

Summary

The biological responses observed in trypanosome-infected mice also occur in cattle, albeit to a much lesser degree. While the correct antibody response and a reduction of immune suppression are critical for resistance in mice, it is the capacity to control anemia that appears to be the more important factor for the determination of survival and productivity in cattle. Differences in the levels of antibodies (IgG1), early complement components and cytokines (IL-4 mRNA) have been found between trypanotolerant and susceptible cattle, but their contribution to differences in control of parasitemia and anemia in these breeds is unknown. Resolution of this requires *in vivo* depletion studies to determine cause and affect relationships.

IMMUNOMODULATED CATTLE
Bovine chimeras combining resistant and susceptible genotypes

It is well established that bovine twin fetuses exchange haemopoietic stem cells through the placentae in early foetal life (Owen, 1945), resulting in haemopoietic chimaerism with sharing of erythrocyte and leukocyte genotypes. Each calf, while genetically different from its twin, carries haemopoietic cells with its own and its twins' genotype.

The responses of bovine chimeras combining resistant and susceptible genotypes were analysed during infection, to obtain information on the role of haemopoietic tissue in trypanotolerance. Five pairs of trypanotolerant N'Dama/susceptible Boran chimaeric cattle (Fig 1) were produced by artificial implantation of a Boran and a N'Dama embryo in the same recipient dam (Naessens et al. submitted). By monitoring the percentages of lymphocytes expressing alleles of the CD5 leukocyte marker, it was established that Boran and N'Dama twins shared haemopoietic cells. The twin calves carried more lymphocytes of Boran origin, ranging from 70 to 95 percent, probably because Boran haemopoietic precursors migrated at an earlier time during fetal development. Furthermore, this percentage was identical in each twin pair and did not change significantly with time. Thus an N'Dama chimeric twin differed from an N'Dama singleton control calf in its haemopietic tissues, which were almost entirely of the susceptible Boran genotype.

Figure 1: Cow with chimeric twins produced by simultaneous implantation of a N'Dama (calf on the left) and a Boran (calf on the right) embryo. These cattle twins are genetically different, but share haemopoietic cells of both genotypes

The twins calves and matched N'Dama and Boran singleton controls were infected with a *T. congolense* clone by tsetse fly challenge. As expected, the N'Dama control animals performed better than the Boran controls after *T. congolense* infection in every aspect: lower parasitemia, higher PCV and more weight gain. However, the N'Dama twins performed in between: they controlled parasitemia as well as the N'Dama singletons, but their PCV decreased like the PVC of the Boran. Thus, if the haemopoietic tissue of a resistant N'Dama is replaced by that of a susceptible animal, it can still control parasitemia, but no longer anemia.

These results show that trypanotolerance is a trait that involves two independent mechanisms: a better capacity to control parasitemia, which is affected by N'Dama genes that are expressed outside of the hemopoietic system, and a better capacity to control anemia, which is a function of N'Dama genes expressed in haemopoietic tissue. The capacity to gain weight during infection was correlated with the capacity to control anemia: weight gain in N'Dama twins was as poor as in Boran, suggesting that there is a relationship between anemia and performance.

T cell depletions

It is possible to *in vivo* deplete T-lymphocytes in cattle by inoculation of monoclonal antibodies to T-lymphocyte antigens (Naessens et al. 1998). Whereas depletion of CD8[+] T-cells in mice reduced parasitemia (Bakhiet et al. 1990), no change in parasitemia or anemia was noted when cattle were depleted for CD8[+] T cells (Sileghem and Naessens, 1995), or both γ,δ- and CD8[+] T cells (De

Buysscher and Naessens, unpublished) using antibodies to CD8[+] and BoWC1 (Naessens et al. 1997), suggesting that these cells do not contribute to the trypanosome's survival in cattle.

We have also depleted N'Dama and Boran cattle of CD4+ T cells (Naessens et al. in preparation). The CD4[+]-depletion drastically reduced the titres of antibodies, whether specific for variant antigens, non-variant antigens or non-trypanosome antigens, in both cattle breeds. No significant difference was observed between parasitemias of CD4[+]-depleted and control N'Dama. In contrast, parasitemias were significantly higher in CD4[+]-depleted Boran cattle than in the control Boran cattle. Thus, CD4[+] T lymphocytes and dependent cells contribute to trypanosome control in susceptible cattle, but not, or to a lesser extent, in tolerant cattle. These data are consistent with the possibility that N'Dama cattle have an innate response that contributes to control of parasitemia and that is absent in Boran cattle. The chimera experiment, discussed above would suggest that this putative innate mechanism is mounted by non-haemopoietic tissues.

Blood packed cell volumes were not different between CD4+ T cell-depleted cattle and controls, whether trypanotolerant or susceptible, suggesting that the T cells and dependent cells do not contribute significantly to the development of anemia.

The biggest surprise of these experiments was that an innate mechanism to control trypanosome parasitemia exists in N'Dama cattle and remains fully functional under conditions (CD4 T cell depletion) that severely impair trypanosome-induced antibody responses. The nature of the innate mechanism is unknown. One possible mechanism described in other species is a trypanolytic serum factor, two of which are found in human sera which has selective lytic activity against *T.b. brucei* (Portela et al. 2000). In addition, high levels of xanthine oxidase in serum from some wild ruminants help control all species of trypanosomes during the acute phase of infection (Black et al. 2001) and high levels of TNF-α are involved in growth control of *T. brucei* in mice (Magez et al. 1997). Another serum factor, such as polyamine oxidase (Traore-Leroux et al. 1987), may be active in trypanotolerant cattle.

MAPPING TRYPANOTOLERANCE GENES

Crosses between resistant and susceptible cattle show intermediate susceptibility, and the trait segregates in a complex manner in the F2 generation suggesting that trypanotolerance is a multigenic trait. The identification of such "resistance" genes would help decipher the molecular mechanisms involved in trypanotolerance. A gene mapping approach was initiated on F2 crosses between N'Dama and Boran cattle (Kemp and Teale, 1998). A total of 200 F2 animals were challenged with *T. congolense*, and the three generations of cattle were genotyped on the basis of a microsatelite-based gene map. A number of quantitative trait loci (QTL) have been identified, but they span a very large area of the chromosome. Efforts are underway to narrow these areas, and to identify possible "trypanotolerance" genes through comparative mapping with the human genome and with a mouse model for trypanotolerance (Teale et al. 1999; Nillson et al. 1999). However, it is unclear whether the mice have the same resistance mechanisms as N'Dama cattle and consequently, the efficacy of this approach remains uncertain.

CONCLUSIONS

There is now ample evidence that trypanotolerance in cattle encompasses at least two mechanisms. One confers a better capacity to control parasitemia, the other a better capacity to control anemia. From our chimeric and T cell depletion studies, it appears that efficient control of parasitemia in trypanotolerant cattle is linked to non-hemopoietic tissues and is little affected by manipulations that block T cell responses and antibody production. This stands in contrast to studies in trypanosusceptible breeds of cattle and in mice where such manipulations lead to shorter survival times and higher parasitemias. The capacity of N'Dama to control anemia is due to genes expressed in hemopoietic tissues, but is not affected by manipulations that inhibit T cell and antibody responses. This suggests a lack of involvement of the acquired immune response in resistance to anemia. Resistance to anemia in trypanosome-infected N'Dama may involve lower macrophage activation and red cell modification for phagocytosis, or an enhanced haemopoietic response, or combinations of these. It is likely that such a mechanism may also be advantageous in controlling other diseases that induce anemia. In this regard, trypanotolerant Red Masai sheep display a reduced level of anemia in response to trypanosome infection than the more susceptible Merino sheep (Murray et al. 1982). They are also more resistant to intestinal nematodes, that induce severe anemia in susceptible animals (Baker et al. 1999).

So what does the future hold? New genes responsible for traits that are associated with "trypanotolerance" may be identified using new and more powerful mapping techniques. While the discovery of new genetic markers may not, in itself, explain all the complexities of host-parasite relations, the identification of new genes will undoubtedly contribute to an understanding of the molecular pathways that lead to trypanotolerance. Knowledge of the nature of trypanotolerance will benefit not only the development of strategies to control trypanosomiasis, but may well lead to new insights in the control of anemia and infectious disease in general.

Acknowledgements

I wish to thank the following people for their contributions to this manuscript: Dr. E. Authié, Prof. S.J. Black, Dr. O. Hanotte, Prof. B. Wilkie

REFERENCES

Andrianarivo, A.G., P. Muiya, M. Opollo, and L.L. Logan-Henfrey. 1995. *Trypanosoma congolense*: comparative effects of a primary infection on bone marrow progenitor cells from N'Dama and Boran cattle. Experimental Parasitology **80**: 407-418.

Assoku, R.K.G., and P.R. Gardiner. 1989. Detection of antibodies to platelets and erythrocytes during haemorrhagic *Trypanosoma vivax* infection of Ayrshire cattle. Veterinary Parasitology. 31: 199-216.

Authié, E., and T. Pobel. 1990. Serum haemolytic complement activity and C3 levels in bovine trypanosomosis under natural conditions of challenge – early indications of individual susceptibility to disease. Veterinary Parasitology. **35**: 43-59.

------------, E., D.K. Muteti, Z.R. Mbawa, J.D. Lonsdale-Eccles, P. Webster, and C.W. Wells. 1992. Identification of a 33-kilodalton immunodominant antigen of *Trypanosoma congolense* as a cysteine protease. Molecular and Biochemical Parasitology. **56**: 103-116.

------------, G. Duvallet, C. Robertson, and D.J.L. Williams. 1993. Antibody responses to a 33 kDa cysteine protease of *Trypanosoma congolense*: relationship to 'trypanotolerance' in catlle. Parasite Immunology. **15**: 465-474.

------------, E., D.K. Muteti, and D.J.L. Williams. 1993. Antibody responses to invariant antigens of *Trypanosoma congolense* in cattle of differing susceptibility to trypanosomiasis. Parasite Immunology. **15**: 101-11

Baker, R.L., D.M. Mwamachi, J.O. Audho, E.O. Aduda, and W. Thorpe. 1999. Genetic resistance to gastro-intestinal nematode parasites in Red Maasai, Dorper and Maasai x Dorper ewes in the sub-humid tropics. Animal Science **69**: 335-344

Bakhiet, M., T. Olsson, P. Van der Meide, and K. Kristensson. 1990. Depletion of CD8[+] T cells suppresses growth of *Trypanosoma brucei brucei* and interferon-gamma production in mice. Clinical Experimental Immunology, **81**: 195-199.

Beschin, A., L. Brys, S. Magez, M. Radwanska, and P. De Baetselier. 1998. *Trypanosoma brucei* infection elicits nitric oxide-dependent and nitric oxide-independent suppressive mechanisms. Journal of Leukocyte Biology. **63**: 429-439.

Buza, J.J., M. Sileghem, P. Gwakisa, and J. Naessens. 1997. CD5[+] B lymphocytes are the main source of antibodies reactive with non-parasite antigens in *Trypanosoma congolense*-infected cattle. Immunology, **92**: 226-233.

------------, and J. Naessens. 1999. Trypanosome non-specific IgM antibodies detected in serum of *Trypanosoma congolense*-infected cattle are polyreactive. Veterinary Immunology and Immunopathology, **69**: 1-9.

Crowe, J.S., A.G. Lamont, J.D. Barry, and K. Vickerman. 1984. Cytotoxicity of monoclonal antibodies to *Trypanosoma brucei*. Transactions of the Royal Society of Tropical Medicine and Hygiene. **78**: 508-513.

108   *Naessens et al.*

Dargie, J. D., P.K. Murray, M. Murray, W.R.I. Grimshaw, and W.I.M. McIntyre. 1979. Bovine trypanosomiasis: the red cell kinetics of N'Dama and Zebu cattle infected with *Trypanosoma congolense*. Parasitology **78**: 271-286.

Darji, A., A. Beschin, M. Sileghem, H. Heremans, L. Brys, and P. De Baetselier. 1996. *In vitro* simulation of immunosuppression caused by *Trypanosoma brucei*: active involvement of gamma interferon and tumor necrosis factor in the pathway of suppression. Infection and Immunity. **64**: 1937-1943.

Ellis, J.A., J.R. Scott, N.D. Machugh, G. Gettingby, and W.C. Davis. 1986. Peripheral blood leucocytes subpopulation dynamics during *Trypanosoma congolense* infection in Boran and N'Dama cattle: an analysis using monoclonal antibodies and flow cytometry. Parasite Immunology. **9**: 363-378.

Epstein, H., and I.L. Mason. 1983. Cattle. *In* Evolution of domesticated animals. I.L. Mason (ed.). Longman, London & New York, p. 6-27.

Flynn, J.N., and M. Sileghem. 1993. Immunosuppression in trypanotolerant N'Dama cattle following *Trypanosoma congolense* infection. Parasite Immunology.**15**: 547-552.

Goerdt, S., and C.E. Orfanos. 1999. Other functions, other genes: alternative activation of antigen-presenting cells. Immunity. **10**:137-42.

Griffin, L., and E.W. Allonby 1979. Trypanotolerance in breeds of sheep and goat with an experimental infection of *Trypanosoma congolense*. Veterinary Parasitology. **5**: 975-*

Hanotte, O., C.L. Tawah, D.G. Bradley, M. Okomo, Y. Verjee, J. Ochieng, and J.E.O. Rege. 2000. Geographic distribution and frequency of a taurine *Bos taurus* and an indicine *Bos indicus* Y specific allele amongst sub-Saharan African cattle breeds. Molecular Ecology. **9**: 387-396.

Kemp S.J., and A.J. Teale. 1998. Genetic basis of trypanotolerance in cattle and mice. Parasitology Today, **14**: 450-454.

Luckins, A.G. 1976. The immune response of zebu cattle to infection with *Trypanosoma congolense* and *T. vivax*. Annals of Tropical Medicine and Parasitology. **70**: 133-145.

Magez, S., M. Geuskens, A. Beschin, H. del Favero, H. Verschueren, R. Lucas, E. Pays, and P. de Baetselier. 1997. Specific uptake of tumor necrosis factor-alpha is involved in growth control of *Trypanosoma brucei*. Journal Cellular Biology. **137**: 715-727.

------------, B. Stijlemans, M. Radwanska, E. Pays, M.A.J. Ferguson and P. De Baetselier. 1998. The glycosyl-inositol-phosphate and dimyristoylglycerol moieties of the glycosylphosphatidylinositol anchor of the Trypanosome variant-specific surface glycoprotein are distinct macrophage-activating factors. Journal of Immunology, **160**: 1949-1956.

Mbawa, Z.R., I.D. Gumm, E. Shaw, and J.D. Lonsdale-Eccles. 1992. Characterisation of a cysteine protease from bloodstream forms of *Trypanosoma congolense*. European Journal of Biochemistry. **204**: 371-379.

Mertens, B., K. Taylor, C. Muriuki, and M. Rocchi. 1999. Cytokine mRNA profiles in trypanotolerant and trypanosusceptible cattle infected with the protozoan parasite *Trypanosoma congolense*: protective role for interleukin-4? Journal of Interferon and Cytokine Research. **19**: 59-65.

Muranjan, M., Q. Wang, Y.L. Li, E. Hamilton, F.P. Otieno-Omondi, J. Wang, A. Van Praagh, J.G. Grootenhuis, and S.J. Black. 1997. The trypanocidal Cape buffalo serum protein is xanthine oxidase. Infection and Immunity **65**: 3806-3814.

Murray, M., W.I. Morrison, and D.D. Whitelaw. 1982. Host susceptibility to African trypanosomiasis: Trypanotolerance. Advances in Parasitology. **21**: 1-68.

--------------, and T.M. Dexter. 1988. Anaemia in bovine African trypanosomiasis. Acta Tropica. **45**: 389-432.

Musoke, A.J., V.M. Nantulya, A.F. Barbet, F. Kironde, and T.C. McGuire. 1981. Bovine immune response to African trypanosomes, specific antibodies to variable surface glycoproteins of *Trypanosoma brucei*. Parasite Immunology. **3**: 97-106.

Mutayoba, B.M., S. Gombe, E.N. Waindi, and G.P. Kaaya. 1989. Comparative trypanotolerance of the small East African breed of goats from different localities to *Trypanosoma congolense* infection. Veterinary Parasitology. 31: 95-*

Naessens, J., and D.J.L. Williams. 1992. Characterization and measurement of CD5[+] B cells in normal and *Trypanosoma congolense*-infected cattle. European Journal of Immunology. **22**: 1713-1718.

--------------, C.J. Howard, and J. Hopkins. 1997. Nomenclature and characterisation of leukocyte differentiation antigens in ruminants. Immunology Today. **18**: 365-368.

--------------, J., J.P. Scheerlinck, E.V. De Buysscher, D. Kennedy, and M. Sileghem. 1998. Total depletion of T cell subpopulations and loss of memory in cattle using mouse monoclonal antibodies. Veterinary Immunology and Immunopathology. **64**: 219-234.

--------------, J., S.G.A. Leak, D.J. Kennedy, S.J. Kemp, and S.J. Teale. 2001. Responses of bovine chimaeras combining trypanosomiasis resistant and susceptible genotypes to experimental infection with *Trypanosoma congolense*. Submitted.

Ngaira, J.M., V.M. Nantulya, A.J. Musoke, and K. Hirumi. 1983. Phagocytosis of antibody-sensitized *trypanosoma brucei in vitro* by bovine peripheral blood monocytes. Immunology. **49**: 393-400.

Nielsen K., J. Sheppard, W. Holmes, and I. Tizard. 1978. Experimental bovine trypanosomiasis: changes in the catabolism of serum immunoglobulins and complement components in infected cattle. Immunology. **35**: 817-826.

Nilsson P.., S. Kang'a, K. Rottengatter, U. Suebeck, F. Iraqi, J. Mwakaya, D. Mwangi, J.E. Womack, T. Goldammer, M. Schwerin, D. Bradley, M. Agaba, K. Sugimoto, A. Gelhaus, R. Horstmann, A. Teale, S. Kemp, and O. Hanotte. 1999. Radiation hybrid maps of candidate trypanotolerance regions in cattle. *In* Proceedings of the International Symposium Candidate Genes for Animal Health, 25-27 aout 1999. Arch. Tierz. Dummerstorf 42, Special Issue, Rostock/Germany, p. 123-125.

Ochsenstein, A.F., and R.M. Zinkernagel. 2000. Natural antibodies and complement link innate and acquired immunity. Immunology Today. **21**: 624-630

Owen, R.D. 1945. Immunogenetic consequences of vascular anastomoses between bovine twins. Science. **102**: 400-401.

Paling, R.W., S.K. Moloo, J.R. Scott, F.A. McOdimba, L.L. Logan-Henfrey, M. Murray, and D.J. Williams. 1991. Susceptibility of N'Dama and Boran cattle to tsetse-transmitted primary and rechallenge infections with ahomologous serodeme of *Trypanosoma congolense*. Parasite Immunology. **13**: 413-425.

--------------, S.K. Moloo, J.R. Scott, G. Gettinby, F.A. McOdimba, and M. Murray. 1991. Susceptibility of N'Dama and Boran cattle to sequential challenges with tsetse-transmitted clones of *Trypanosoma congolense*. Parasite Immunology. **13**: 427-45.

Pinder, M., F. Fumoux, A. van Melick, and G.E. Roelants. 1987. The role of antibody in natural resistance to African trypanosomiasis. Veterinary Immunology and Immunopathology. **17**:325-32.

Portela, M., J. Raper, and S. Tomlinson. 2000. An investigation in the mechanism of trypanosome lysis by human serum factors. Molecular and Biochemical Parasitology. **110**: 273-282.

Russo, D.C.W., D.J.L. Williams, and D.J. Grab. 1994. Mechanisms for the elimination of potentially lytic complement-fixing variable surface glycoprotein antibody-complexes in *Trypansoma brucei*. Parasitology Research. **80**: 487-492.

Sileghem, M., and J.N. Flynn. 1992. Suppression of interleukin 2 secretion and interleukin 2 receptor expression during tsetse-transmitted trypanosomiasis in cattle. European. Journal of Immunology. **22**: 767-773.

----------------, J.N. Flynn, R. Saya, and D.J.L. Williams. 1993. Secretion of costimulatory cytokines by monocytes and macrophages during infection with *Trypanosoma (Nannomonas) congolense* in susceptible and tolerant cattle. Veterinary Immunology and Immunopathology. **37**: 123-134.

----------------, A. Darji, P. De Baetselier, J.N. Flynn, and J. Naessens. 1994. African Trypanosomiasis. *In* Parasitic infections and the immune system. F. Kierszenbaum (ed.). Academic Press., San Diego, CA, p. 1-51

----------------, J.N. Flynn, L.L. Logan-Henfrey, and J.A. Ellis. 1994. Tumor necrosis factor a production by monocytes from cattle infected with *Trypanosoma (Duttonella) vivax* and *Trypanosoma (Nannomonas) congolense:* Correlation with severity of anemia associated with the disease. Parasite Immunology. **16**: 51-54.

----------------, and J. Naessens. 1995. Are CD8 T cells involved in control of African trypanosomiasis in a natural host environment? European Journal of Immunology. **25**: 1965-1971.

----------------, R. Saya, D.J. Grab, and J. Naessens. 2001. An accessory role for the diacylglycerol moiety of variable surface glycoprotein of African trypanosomes in the stimulation of bovine monocytes. Veterinary Immunology and Immunopathology, 78, 325-339.

Sternberg, J., and McGuigan F. 1992. Nitric oxide mediates suppression of T-cell responses in murine *Trypanosoma brucei* infection. European Journal of Immunology. **22**: 2741-2744.

Tabel, H., G.J. Losos, and M.G. Maxie. 1980. Experimental bovine trypanosomiasis (*Trypanosoma vivax* and *T. congolense*). II. Serum levels of total protein, albumin, haemolytic complement, and complement component C3. Tropenmedizin und Parasitologie. **32**: 149-153.

Taylor, K.A., V. Lutje, and B. Mertens. 1996. Nitric oxide synthesis is depressed in *Bos indicus* cattle infected with *Trypanosoma congolense* and *T. vivax* and does mediate T cell suppression. Infection and Immunity. **64**: 4115- 4122.

----------------, V. Lutje, D. Kennedy, E. Authié, A. Boulange, L.L. Logan-Henfrey, B. Gichuki, and G. Gettinby. 1996. *Trypanosoma congolense:* B-lymphocyte responses differ between trypanotolerant and trypanosusceptible cattle. Experimental Parasitology. **83**: 106-116.

----------------, B. Mertens, V. Lutje, and R. Saya 1998. *Trypanosoma congolense* infection of trypanorolerant N'Dama *(Bos taunus)* cattle is associated with decreased secretion of nitric oxide by interferon-γ-activated monocytes and increased transcription of interleukin-10. Parasite Immunology. **20**: 421-429.

Teale A., M. Agaba, S.T. Clapcott, A. Gelhaus, C.H. Haley, O. Hanotte, R. Horstmann, F. Iraqi, S.T. Kemp, P. Nilsson, M. Schwerin, K. Sekikawa, M. Soller, Y. Sugimoto, and J. Womack. 1999. Resistance to trypanosomosis: of markers, genes and mechanisms. *In* Proceedings of the International Symposium Candidate Genes for Animal Health, 25-27 aout 1999. Arch. Tierz. Dummerstorf 42, Special Issue, Rostock/Germany, p. 36-41.

Toure, S.M., M. Seye, T. Dieye, and M. Mbengue 1983. Trypanotolerance-comparative pathological survey in Djallonke and Fulane sheep of the Sahel. *In* International Scientific Council for Trypanosomiasis Research and Control, 17th meeting, Arusha, Tanzania, 1981. Organization of African Unity OAU/STRC, publication nr 112. p326.

Traore-Leroux, T., F. Fumoux , J. Chaize, and G.E. Roelants. 1987. *Trypanosoma brucei:* polyamine oxidase mediated trypanolytic activity in the serum of naturally resistant cattle. Experimental Parasitology. **64**: 401-409.

Trail J.C.M., G.D.M. d'Ieteren, A. Feron, O. Kakiese, M. Mulungo, and M. Pelo. 1991. Effect of trypanosome infection, control of parasitaemia and control of anaemia development on productivity iof N'Dama cattle. Acta Tropica. **48**: 37-45.

--------------, G.D.M. d'Ieteren, J.C. Maille, and G. Yangari. 1991. Genetic aspects of control of anaemia development in trypanotolerant N'Dama cattle. Acta Tropica. **48**: 285-291.

Wei, G., L. Qualtiere, and H. Tabel. 1990. *Trypanosoma congolense*: complement-independent immobilization by a monoclonal antibody. Experimental Parasitology. **70**: 483-485.

Williams, D.J.L., J. Naessens, J.R. Scott and F.A. McOdimba. 1991. Analysis of peripheral leucocyte populations in N'Dama and Boran cattle following a rechallenge infection with *Trypanosoma congolense*. Parasite Immunology. **13**: 171-185.

------------------, K. Taylor, J. Newson., B. Gichuki, and J. Naessens. 1996. The role of variable surface glycoprotein antibody responses in bovine trypanotolerance. Parasite Immunology. **18**: 209-218.

TRYPANOSOME FACTORS CONTROLLING POPULATION SIZE AND DIFFERENTIATION STATUS

N.B. Murphy[1], and T. Olijhoek[2]

[1]International Livestock Research Institute, P.O. Box 30709, Nairobi, Kenya.
[2]ID-Lelystad, Insitute for Animal Science and Health, Edelhertweg 15, 8219 PH Lelystad, The Netherlands.

ABSTRACT

African trypanosomes undergo a complex life cycle involving an insect vector and vertebrate host. Growth, differentiation and population size in both the vector and host species are carefully controlled, ensuring survival and transmission of the parasite. Population control in vector and mammal hosts is associated with parasite transition from replicating to non-replicating forms, the latter of which is specialized for infection. Recently it was shown that bloodstream form trypanosomes release a low molecular weight factor that feeds back on the parasites causing them to cease division. We have found that the factor acts between the different African trypanosome species, consistent with a common mechanism of growth control. The factor also acts on tsetse-transmitted metacyclic form trypanosomes blocking their infectivity for mammals. Our data suggests that the factor is internalized via a trypanosome cell surface receptor, binds to adenosine ribosylation factor 1 and suppresses endocytosis.

Key words Quorum sensing; ADP ribosylation factor; slender; stumpy; trypanosome; cyclic AMP; adenylate cyclase; density-dependent control; differentiation; receptor.

THE LIFE CYCLE

The life cycle of African trypanosomes is complex and alters between the tsetse fly vector and vertebrate host (Vickerman, 1985; Van Den Abbeele et al., 1999). The classical life cycle described in textbooks is that of *Trypanosoma brucei brucei*. (Fig. 1). Other species of African trypanosomes differ mainly in their site of development within the tsetse fly vector and some in their mode of transmission. At different stages of their life cycle *T. brucei* alternates between proliferative (slender, procyclic, epimastigote) and quiescent (stumpy, proventricular, metacyclic) developmental stages

(Vickerman, 1985; Matthews and Gull, 1994; Mottram, 1994) each of which has one or more unique morphological features (Fig 1) and physiological properties. We are studying the differentiation of bloodstream stage *T. brucei* from replicating slender to non-replicating stumpy forms because this transition heralds the acquisition of tsetse infectivity and loss of growth potential in the mammal (Balber 1972; Black et al., 1985; Seed and Secheklski, 1988). We anticipate that identification of the molecular triggers for the slender to stumpy transition will suggest ways to prevent growth of African trypanosomes in mammals.

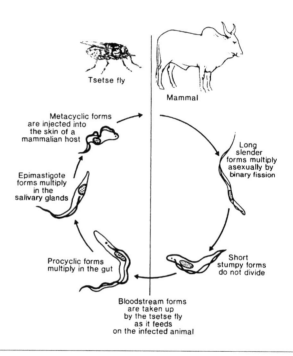

Figure 1. *Simplified life cycle of trypanosomes showing the principal stages undergoing cyclical transmission between a cattle host and tsetse fly vector.*

WHAT TRIGGERS DIFFERENTIATION?

Differentiation of slender to stumpy form *T. brucei* occurs in immunosuppressed animals excluding a requirement for components of acquired immune responses. Population transition occurs at or slightly before peak parasitemia which led investigators to propose that it might be density-dependent (Black et al., 1985; Seed and Sechelski, 1989; Hamm et al., 1990), and subsequently to the development of a mathematical model based on the premise that slender form organisms release a stumpy-inducing factor (Seed and Black, 1997, 1999). Studies with conditioned media has firmly established that bloodstream stage *T. brucei* controls its growth and differentiation through a parasite-produced factor *in* vitro (Hesse et al., 1995; Selzer et al., 1996; Vassella et al., 1997).

The interplay between cell cycle progression and differentiation of bloodstream form trypanosomes has also received attention (Matthews and Gull, 1994; Matthews 1999). Studies at the individual cell level have revealed the existence of a differentiation division in bloodstream trypanosomes involving a committed slender cell dividing to produce daughter cells, at least one of which undergoes a morphological transformation without further division (Tyler et al., 1997). Based on the observation that stumpy cells are uniformly arrested in their cell cycle and transform synchronously, a model that describes the initiation of differentiation as being cell cycle position dependent has been developed (Ziegelbauer et al., 1990; Matthews and Gull 1994). Specifically, cells in the G1–G0 phase of the cell cycle would be competent to receive the differentiation signal, whereas those outside this window would not.

INTER-SPECIES AND LIFE CYCLE STAGE SIGNALS

Axenic culture systems are available for several species of African trypanosomes (Hirumi and Hirumi, 1991). We used these to evaluate whether species other than *T. brucei* generate trypanosome growth-inhibitory materials. Not only did we find that parasite species have similar density controlling mechanisms to *T. brucei*, but the parasite-released factors from any one species can induce growth arrest and differentiation of the other species of trypanosomes (Olijhoek and Murphy, unpublished). This is analogous to quorum sensing in bacteria in which acyl-homoserine lactones that regulate growth control within a species also allow bacterial populations to interact with each other as well as with their eukaryotic hosts (Eberl, 1999). Recently, studies on the dynamics of multiple infections of malaria parasites in asymptomatic children living under intense transmission pressure also provide evidence for a density-dependent regulation that transcends species as well as genotype (Bruce et al., 2000).

The growth of *T. congolense* procyclic and epimastigote forms *in vitro* was not affected by the presence of conditioned medium. The same medium inhibited the conversion of metacyclic to bloodstream stage *T. congolense* but this inhibition was reversible for up to 120 hours of incubation. The conditioned medium also stopped replication of bloodstream stage *T. congolense* but the growth inhibition was irreversible after 12 hours and the parasites appeared to undergo a form of cell death in the ensuing 24 hours. Thus the bloodstream stage trypanosomes release a factor, or factors, that prevent(s) differentiation of metacyclic to bloodstream stage organisms, and cause(s) irreversible growth-inhibition of the latter.

PARASITE FACTORS IN POPULATION CONTROL

Conditioned media of bloodstream form trypanosomes that contains the growth-inhibitory factors elicits an immediate two- to threefold elevation of intracellular cAMP upon addition to slender *T. brucei* (Vassella et al., 1997). This is consistent with estimates of cAMP levels of trypanosomes in rats in which a three-fold increase in cAMP levels is observed at peak parasitaemia and then decreases during the transition to stumpy forms (Mancini and Patton, 1981). Membrane-permeable derivates of cAMP or the phosphodiesterase inhibitor etazolate perfectly mimic the differentiation factor activity indicating that the cell cycle regulation of bloodstream forms is under dominant control of cAMP signalling involving adenylate cyclase (Vassella et al., 1997).

Differentiation of a related organism, *Trypanosoma cruzi*, from the proliferating epimastigote to the quiescent metacyclic stage has also been correlated to activation of adenylate cyclase (Gonzales-Perdomo et al., 1988; Rangel-Aldao et al., 1988; Fraidenraich et al., 1993). The increase in cAMP and growth inhibition is induced by an $\alpha(d)$-globin fragment from the *Triatoma* insect vector hindgut (Fraidenraich et al., 1993) and is a mediated by a cAMP-dependent protein kinase (Malaquias and Oliveira, 1999). A link between cAMP signaling and the activity of protein kinases identified in African trypanosomes has yet to be established.

Quorum sensing in *Vibrio fischeri* via an acyl-homoserine lactone also involves cAMP and the cAMP receptor protein of this bacterium (Dunlap, 1999). This mechanism of population density control may turn out to be universal for gram-negative bacteria subject to quorum sensing via acyl-homoserine lactones.

OTHER EVENTS INDUCING STUMPY FORMS

We have examined endocytosis and vesicle formation following the treatment of slender form *T. brucei* with conditioned media and found that vesicles rapidly reduce in size and number and are remarkably uniform in comparison to untreated cells, at least when the slender forms are from populations of pleomorphic *T. brucei* which are capable of the slender to stumpy differentiation.

Endocytosis in these cells is essentially blocked (Murphy and Olijhoek, unpublished). In contrast, populations of slender *T. brucei* That have lost the capacity to differentiate to stumpy forms, called monomorphic *T. brucei*, are unaffected.

The availability of the trypanosome ADP ribosylation factor 1 (*T-ARF1*) gene in our laboratory (Osanya, 1999) encoding a ubiquitous and essential protein involved in vesicle formation, membrane trafficking and signaling (Chavrier and Goud, 1999; Randazzo et al., 2000), prompted us to examine whether this protein may be a target for material in the trypanosome growth-inhibitory medium. Surprisingly when conditioned medium was passed through a column containing recombinant T-ARF1, the trypanosome- growth inhibitory factor (T-GIF) was removed. The binding of T-GIF to T-ARF1 is very strong and it binds to the denatured as well as native form of the protein.

T-ARF1 is an internal protein and, if it is the natural ligand for T-GIF, a surface receptor must be implicated in the transport of T-GIF across the plasma membrane. Recently, Nolan et al. (1999) established that surface-expressed proteins of trypanosomes contain linear chains of poly-*N*-acetyllactosamine and can be isolated by tomato lectin. Pleomorphic trypanosomes contain a highly abundant

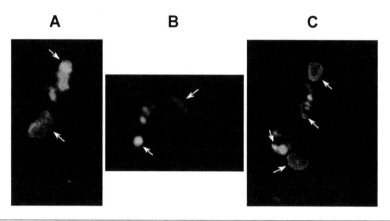

Figure 2. *Localisation of (A) T-ARF1 and tomato lectin-binding proteins in* T. brucei *(B) monomorphic and (C) pleomorphic bloodstream forms. The arrows point to the nuclear and kinetoplast DNA. The kinetoplast co-localizes with the parasite flagellar pocket, the principal site of endo- and exocytosis. Note the labelling of the flagellum in panels A and C and the absence of this labelling for the monomorphic parasite in panel B. The other labelling observed in all panels is of the parasite endosomal system.*

tomato lectin-binding protein of about 68 kDa, whereas monomorphic trypanosomes have low levels of this protein (Nolan and Murphy, unpublished). We have called this protein TPP68 for trypanosome pleomorphic protein of 68 kDa. This protein can be surface labeled on intact trypanosomes and also binds to T-ARF1 (Murphy and Olijhoek, unpublished). In addition, binding of TPP68 to T-ARF1 is blocked when either protein is pre-treated with T-GIF. These results suggest that TPP68 may be the receptor for T-GIF and that its reduced levels in monomorphic trypanosomes could account for their lack of sensitivity to T-GIF.

Antisera raised against T-ARF1 recognized the trypanosomal endosomal system, as might be expected from its role in vesicle formation, but unexpectedly also binds along the parasite flagellum (Figure 2A). Fluorescently labeled tomato lectin binds to the flagellar pocket and endosomal system of monomorphic trypanosomes (Figure 2B, Nolan et al., 1999) but for pleomorphic trypanosomes there is also labeling along the flagellum (Figure 2C). Since the principal difference in tomato lectin-binding proteins in pleomorphic and monomorphic trypanosomes is the abundance of TPP68 and since this protein binds to T-ARF1, the labeling of the flagellum of pleomorphic trypanosomes by anti-T-ARF1 mouse sera and by fluorescent tomato lectin suggests that these proteins co-localize in intact pleomorphic trypanosomes. These data are consistent with the possibility that TPP68 is the receptor and T-ARF1 the principal target for T-GIF. Interestingly, these results link back to adenylate cyclase and cAMP, since adenylate cyclase has also been localized along the flagellum of *T. brucei* trypanosomes (Paindavoine et al., 1992). Whether T-ARF1, TPP69 or T-GIF interact directly or indirectly with adenylate cyclase localized along the flagellum remains to be determined

WHY TARGET T-ARF1?

ARF was first identified in animal cells as the cofactor required for the in vitro ADP-ribosylation of the stimulatory regulatory subunit of adenylate cyclase, G_s, by cholera toxin. The relevance of this activity to *in vivo* function is unknown, but recently it was shown that that the interaction between ARF and cholera toxin is highly specific and is required *in vivo* for maximal toxicity (Jobling and Holmes, 2000). Studies on the ARF family of small GTP binding proteins have shown these to be essential, conserved, multi-functional proteins involved in the regulation of vesicle formation, exo- and endocytosis, membrane recycling, the actin cytoskeleton and molecular switches in various signal transduction pathways (Chavrier and Goud, 1999; Randazzo et al., 2000). ARF in combination with members of the protein kinase C family and of the Rho subfamily of small GTPases influence the activity of phospholipase D (PLD). Since the ARF subfamily of small GTPases are potent activators of PLD, this suggests that PLD is the focal point for integration of cellular responses to signaling and homeostasis (Cockroft, 1996; Ktistakis, 1998). PLD activation has direct effects on regulated exocytosis, actin stress fiber formation, cell morphology and motility (Du et al., 2000) and these functions fit well with the morphological changes observed in trypanosomes in their transition from long slender to short stumpy forms.

THE STORY SO FAR

Trypanosome bloodstream form-produced factors induce growth arrest and differentiation of long slender to short stumpy forms that undergo programmed cell death, unless transferred to conditions that allow differentiation to procyclic forms. The factors operate between the different trypanosome species and influence the establishment of infection by metacyclic forms from tsetse flies. The factor, named T-GIF, binds to a 68 kDa receptor of *T. brucei*, TPP68, that co-localizes with T-ARF1 and possibly adenylate cyclase. Whether adenylate cyclase also associates with TPP68 and/ or T-ARF1 is not known, but it is involved in the slender to stumpy switch in altering the levels of cAMP (Vassella et al., 1997). The localization of each of these proteins involved in this signal transduction pathway is unlikely to be due to chance or coincidence. Following binding to TPP68, T-GIF is transferred to T-ARF1, through a process that is yet to be determined, and then dissociates from TPP68. The binding of T-GIF to T-ARF1 is irreversible and thus prevents T-ARF1 from associating with other proteins involved in vesicle formation and trafficking. The TPP68 receptors from which T-ARF1 has dissociated can now recruit free T-ARF1 and repeat the cycle of transferring T-GIF to T-ARF1. In this manner the T-ARF1 is not available for vesicle formation and trafficking and for interaction with PLD thus shutting down exo- and endocytosis and inducing morphological and biochemical changes. The effects would be similar to the action of ceramides on human granulocytes through inhibition of PLD activation (Abousalham et al., 1997). There are still many gaps in this outline sketch of cell cycle arrest and differentiation, particularly the link with adenylate cyclase and the role of cAMP in combination with downstream effectors such as protein kinases.

EXPLOITING THE FINDINGS

In rational drug development against pathogens the fundamental steps are to identify a key pathogen protein target that differs to the host, to generate compounds that inactivate the function of that target and to ensure that that compounds can cross the plasma membrane and be delivered intact to the target. Each of these represents a formidable and costly problem and funding is not available to overcome these for diseases of the Worlds poorest people. Elucidation of the structure of T-GIF, the surface receptor to which it binds and its signaling pathway in trypanosomes may offer solutions to these problems in drug design, particularly since this factor acts across the different species of trypanosomes and on the infective metacyclic as well as bloodstream forms.

REFERENCES

Abousalham, A. C. Liossis, L. O'Brien, and D.N. Brindley. 1997. Cell-permeable ceramides prevent the activation of phospholipase D by ADP-ribosylation factor and RhoA. Journal of Biol. Chem. **272**: 1069-1075.
Aktories, K. 1997. Rho proteins: targets for bacterial toxins. Trends Microbiol. **5**: 282-288.

Alexandre, S. P. Paindavoine, F. Hanocq-Quertier, F. Paturiaux-Hanocq, P. Tebabi, and E. Pays. 1996. Families of denylate cyclase genes in *Trypanosoma brucei*. Molecular and Biochemical Parasitology **77**: 173-182.

Ashcroft, M.T. 1960. A comparison between a syringe passaged and a tsetse fly transmitted line of a strain of *Trypanosoma rhodesiense*. Annals of Tropical Medicine and Parasitology. **54**: 44–70.

Bacchi, C.J., J. Garofalo, D. Mockenhaupt, P.P. McCann, K.A. Diekema, A.E. Pegg, H.C. Nathan, E.A. Mullaney, L. Chunosoff, A. Sjoerdsma, and S.H. Hutner. 1983. In vivo effects of alpha-DL-difluoromethylornithine on the metabolism and morphology of *Trypanosoma brucei brucei*. Molecular and Biochemical Parasitology **7**: 209-225.

Balber, A. E. 1972. *Trypanosoma brucei*: fluxes of the morphological variants in intact and X-irradiated mice. Experimental Parasitology **31**: 307-319.

Barry, J.D., and K. Vickerman. 1979. *Trypanosoma brucei*: loss of variable antigens during transformation from bloodstream to procyclic forms in vitro. Experimental Parasitology **48**: 313–324.

Bass, K.E., and C.C. Wang. 1991. The *in vitro* differentiation of pleomorphic *Trypanosoma brucei* from bloodstream into procyclic form requires neither intermediary nor short-stumpy stage. Molecular and Biochemical Parasitology **44**: 261–270.

Bienen, E.J., E. Hammadi, and G.C. Hill. 1981. *Trypanosoma brucei*: biochemical and morphological changes during *in vitro* transformation of bloodstream- to procyclic-trypomastigotes. Experimental Parasitology **51**: 408–417.

Black, S.J., C.N. Sendashonga, C. O'Brien, N.K. Borowy, J. Naessens, P. Webster, and M. Murray. 1985. Regulation of parasitaemia in mice infected with *Trypanosoma brucei*. Current Topics in Microbiology and Immunology 117, 93-118.

Bruce, M.C., C.A. Donnelly, M.P. Alpers, M.R. Galinski, J.W. Barnwell, D. Walliker, and K.P. Day. 2000. Cross-species interactions between malaria parasites in humans. Science **287**: 845-848.

Brun, R., and M. Schonenberger. 1981. Stimulating effect of citrate and *cis*-aconitate on the transformation of *Trypanosoma brucei* bloodstream forms to procyclic forms in vitro. Z. Parasitenkd. **66**: 17–24.

Bulow, R., and P. Overath. 1985. Synthesis of a hydrolase for the membrane form-variant surface glycoprotein is repressed during transformation of *Trypanosoma brucei*. FEBS Letters **187**: 105–110.

Carruthers, V.B., and G.A.M. Cross. 1992. High-efficiency clonal growth of bloodstream- and insect-form *Trypanosoma brucei* on agarose plates. Proceedings of the National Academy of Sciences, USA **89**: 8818-8821.

Chavrier, P., and B. Goud. 1999. The role of ARF and Rab GTPases in membrane transport. Current Opinion in Cell Biology **11**: 466-475.

Cockcroft, S. 1996. Phospholipid signaling in leukocytes. Current Opinion in Hematology **3**: 48-54

Cross, G.A., L.E. Wirtz, and M. Navarro. 1998. Regulation of vsg expression site transcription and switching in *Trypanosoma brucei*. Molecular and Biochemical Parasitology **91**: 77-91.

Czichos, J., C. Nonnengasser, and P. Overath. 1986. *Trypanosoma brucei*: *cis*-aconitate and temperature reduction as triggers of synchronous transformation of bloodstream to procyclic trypomastigotes *in vitro*. Experimental Parasitology **62**: 283–291.

de Gee, A.L., P.H. Carstens, P.P. McCann, and J.M. Mansfield. 1984. Morphological changes in *Trypanosoma brucei rhodesiense* following inhibition of polyamine biosynthesis in vivo. Tissue and Cell **16**: 731-738.

Dirie, M.F., S.L. Croft, and D.H. Molyneux 1986. Morphological changes of *Trypanosoma vivax* in mice. Veterinary Parasitology **19**: 23-27.

Du, G., Y.M. Altshuller, Y. Kim, J.M. Han, S.H. Ryu, A.J. Morris, and M.A. Frohman. 2000. Dual requirement for rho and protein kinase C in direct activation of phopholipase D1 through G protein-coupled receptor signalling. Molecular Biology of the Cell **11**: 4359-4368.

Dunlap, P.V. 1999. Quorum regulation of luminescence in *Vibrio fischeri*. Journal of Molecular Microbiology and Biotechnology **1**: 5-12.

Dwinger, R.H., A.G. Luckins, M. Murray, P. Rae, and S.K. Moloo. 1986. Interference between different serodemes of *Trypanosoma congolense* in the establishment of superinfections in goats following transmission by tsetse. Parasite Immunology **8**: 293-305.

-----------, M. Murray, A.G. Luckins, P.F. Rae, and S.K. Moloo. 1989. Interference in the establishment of tsetse-transmitted *Trypanosoma congolense*, *T. brucei* or *T. vivax* superinfections in goats already infected with *T. congolense* or *T. vivax*. Veterinary Parasitology **30**: 177-189.

Eberl, L. 1999. N-acyl homoserinelactone-mediated gene regulation in gram-negative bacteria. Systems in Applied Microbiology **22**: 493-506.

Estevez, A.M., and L. Simpson. 1999. Uridine insertion/deletion RNA editing in trypanosome mitochondria - a review. Gene **240**: 247-260.

Fairbairn, H., and A. Culwick. 1947. The modification of *Trypanosoma rhodesiense* on prolonged syringe passage. Annals of Tropical Medicine and Parasitology **41**: 26.

Field, M.C., B.R.S. Ali, and H. Field. 1999. GTPases in protozoan parasites: tools for cell biology and chemotherapy. Parasitology Today **15**: 365-371.

Fish, W.R., C.W. Muriuki, A.M. Muthiani, D.J. Grab, and J.D. Lonsdale-Eccles. 1989. Disulfide bond involvement in the maintenance of the cryptic nature of the cross-reacting determinant of metacyclic forms of *Trypanosoma congolense*. Biochemistry **28**: 5415-5421.

Fraidenraich, D., C. Pena, E. L. Isola, E.M. Lammel, O. Coso, A.D. Anel, S. Pongor, F. Baralle, H.N. Torres, and M.M. Flawia. 1993. Stimulation of *Trypanosoma cruzi* adenylyl-cyclase by an □(d.-globin fragment from *Triatoma* hindgut - effect on differentiation of epimastigote to trypomastigote forms. Proceedings of the National Academy of Sciences USA **90**: 10140-10144.

Gonzales-Perdomo, M., P. Romero, and S. Goldenberg. 1988. Cyclic AMP and adenylate cyclase activators stimulate *Trypanosoma cruzi* differentiation. Experimental Parasitology **66**: 205-212.

Grossman, A. D. 1995. Genetic networks controlling the initiation of sporulation and the development of genetic competence in *Bacillus subtilis*. Annual Review of Genetics **29**: 477-508.

Hamm, B., A. Schindler, D. Mecke, and M. Duszenko. 1990. Differentiation of *Trypanosoma brucei* bloodstream trypomastigotes from long slender to short stumpy-like forms in axenic culture. Molecular and Biochemical Parasitology **40**: 13-22.

Hecker, H., P.H. Burri, and S. Bohringer. 1973. Quantitative ultrastructural differences in the mitochondrium of pleomorphic bloodforms of *Trypanosoma brucei*. Experientia **29**: 901-903.

Hesse, F., P.M. Selzer, K. Muhlstadt, and M. Duszenko. 1995. A novel cultivation technique for long-term maintenance of bloodstream form trypanosomes in vitro. Molecular and Biochemical Parasitology **70**: 157-166.

Hirumi, H., and K. Hirumi. 1989. Continuous cultivation of *Trypanosoma brucei* bloodstream forms in a medium containing a low concentration of serum protein without feeder cell layers. Journal of Parasitology **75**: 985-989.

-----------, and K. Hirumi. 1991. *In vitro* cultivation of *Trypanosoma congolense* bloodstream forms in the absence of feeder cell layers. Parasitology **102**: 225-236.

Hursey, B.S. 2001. The programme against African trypanosomiasis: aims, objectives and achievements. Trends in Parasitology **17**: 2-3.

Jobling, J.G., and R.K. Holmes. 2000. Identification of motifs in cholera toxin A1 polypeptide that are required for its interaction with human ADP-ribosylation factor 6 in a bacterial two-hybrid system. Proceedings of the National Academy of Sciences, USA **97**: 14662-14667.

Ktistakis, N.T. 1998. Signalling molecules and the regulation of intracellular transport. Bioessays **20**: 495-504.

Mahan S.M., and S.J. Black. 1989. Differentiation, multiplication and control of bloodstream form *Trypanosoma* (*Duttonella*) *vivax* in mice. Journal of Protozoology **36**: 424-428.

Malaquias, A.T., and M.M. Oliveira. 1999. Phospholipid signalling pathways in *Trypanosoma cruzi* growth control. Acta Tropica **73**: 93-108.

Mancini, P.E., and C.L. Patton. 1981. Cyclic 3´, 5´-adenosine monophosphate levels during the developmental cycle of *Trypanosoma brucei brucei* in the rat. Molecular and Biochemical Parasitology **3**: 19-31.

Matthews, K.R. 1999. Developments in the differentiation of *Trypanosoma brucei*. Parasitology Today 15, 76-80.

-----------, and K. Gull. 1994. Evidence for an interplay between cell cycle progression and the initiation of differentiation between life cycle forms of African trypanosomes. Journal of Cell Biology **125**: 1147–1156.

-----------, and K. Gull. 1998. Identification of stage-regulated and differentiation-enriched transcripts during transformation of the African trypanosome from its bloodstream to procyclic form. Molecular and Biochemical Parasitology **95**: 81-95.

McLintock, L.M.L., C.M.R. Turner, and K. Vickerman. 1990. A comparison of multiplication rates in primary and challenge infections of *Trypanosoma brucei* bloodstream forms. Parasitology **101**: 49-55.

Morrison, W.I., P.W. Wells, S.K. Moloo, J. Paris, and M. Murray. 1982. Interference in the establishment of superinfections with *Trypanosoma congolense* in cattle. Journal of Parasitology **68**: 755-764.

Mottram, J. C. 1994. Cdc2-related protein-kinases and cell-cycle control in trypanosomatids. Parasitology Today **10**: 253-257.

Nantulya, V.M., J.J. Doyle, and Jenni, L. 1978. Studies on *Trypanosoma* (*Nannomonas*) *congolense*. I. On the morphological appearance of the parasite in the mouse. Acta Tropica **35**: 329-337.

Naula, C., and T. Seebeck. 2000. Cyclic AMP signaling in trypanosomatids. Parasitology Today 16, 35-38.

Nilsen, T.W. 1995. Trans-splicing: an update. Molecular and Biochemical Parasitology **73**: 1-6.

Nolan, D.P., M. Geuskens, and E. Pays. 1999. N-linked glycans containing linear poly-N-acetyllactosamine as sorting signals in endocytosis in *Trypanosoma brucei*. Current Biology **9**: 1169-1172.

-----------, S. Rolin, J.R. Rodriguez, J. Van Den Abbeele, and E. Pays. 2000. Slender and stumpy bloodstream forms of *Trypanosoma brucei* display a differential response to extracellular acidic and proteolytic stress. European Journal of Biochemistry **267**: 18-27.

Opperdoes, F.R. 1985. Biochemical peculiarities of trypanosomes, African and South American. British Medical Bulletin **41**: 130-135.

Osanya, A.O. 1999. Identification and characterisation of differentially expressed sequence tags in *Trypanosoma brucei brucei*: Molecular cloning, expression and analysis of ADP ribosylation factor 1 from African trypanosomes. Ph.D. thesis, Department of biology and biochemistry, Brunel University, Uxbridge, U.K.

Overath, P., J. Czichos, U. Stock, and C. Nonnengaesser. 1983. Repression of glycoprotein synthesis and release of surface coat during transformation of *Trypanosoma bruce*i. EMBO Journal **2**: 1721–1728.

----------, J. Czichos, and C. Haas. 1986. The effect of citrate/ *cis*-aconitate on oxidative metabolism during transformation of *Trypanosoma brucei.* European Journal of Biochemistry **160**: 175–182.

Paindavoine, P., S. Rolin, S. Van Assel, M. Geuskens, J.-C. Jauniaux, C. Dinsart, G. Huet, and E. Pays. 1992. A gene from the variant surface glycoprotein expression site encodes one of several transmembrane adenylate cyclases located on the flagellum of *Trypanosoma brucei*. Molecular and Cellular Biology **12**: 1218-1225.

Pardridge, W.M. 1999. Vector-mediated drug delivery to the brain. Advances in Drug Delivery Reviews **36**: 299-321.

Parsons, M., and L. Ruben. 2000. Pathways involved in environmental sensing in trypanosomatids. Parasitology Today **16**: 56-62.

Pays, E., J. Hanocq-Quertier, F. Hanocq, S. Van Assel, D. Nolan, and S. Rolin. 1993. Abrupt RNA changes precede the first cell division during the differentiation of *Trypanosoma brucei* bloodstream forms into procyclic forms *in vitro.* Molecular and Biochemical Parasitology **61**: 107–114.

Penichet, M.L., Y.S. Kang, W.M. Pardridge, S.L. Morrison, and S.U. Shin. 1999. An antibody-avidin fusion protein specific for the transferrin receptor serves as a delivery vehicle for effective brain targeting: initial applications in anti-HIV antisense drug delivery to the brain. Journal of Immunology **163**: 4421-4426.

Randazzo, P.A., Z. Nie, K. Miura, and V.W. Hsu. 2000. Molecular Aspects of the Cellular Activities of ADP-Ribosylation Factors. Science's STKE: http://stke.sciencemag.org/cgi/content/full/OC_sigtrans;2000/59/re1.

Rangel-Aldao, R., F. Triana, V. Fernández, G. Comach, T. Abate, and R. Montoreano. 1988. Cyclic AMP as an inducer of the cell differentiation of *Trypanosoma cruz*i. Biochemistry International **17**: 337-344.

Richardson, J.P., R.P. Beecroft, D.L. Tolson, M.K. Liu, T.W. Pearson. 1988. Procyclin: an unusual immunodominant glycoprotein surface antigen from the procyclic stage of African trypanosomes. Molecular and Biochemical Parasitology **31**: 203-216.

Robertson, M. 1912. Notes on the polymorphism of *Trypanosoma gambiense* in the blood and its relation to the exogenous cycle in *Glossina palpalis*. Proceedings of the Royal Society of London B Biological Sciences **85**: 241–539.

Roditi, I., H. Schwartz, T.W. Pearson, R.P. Beecroft, M.K. Liu, J.P. Richardson, H.J. Buhring, J. Pleiss, R. Bulow, R.O. Williams, and P. Overath. 1989. Procyclin gene expression and loss of the variant surface protein during differentiation of *Trypanosoma bruce*i. Journal of Cell Biology **108**: 737–746.

Rolin, S,. P. Paindavoine, J. Hanocq-Quertier, F. Hanocq, Y. Claes, D. Le Ray, P. Overath, and E. Pays. 1993. Transient adenylate cyclase activation accompanies differentiation of *Trypanosoma brucei* from bloodstream to procyclic forms. Molecular and Biochemical Parasitology **61**: 115–126.

Schroeder, U., H. Schroeder, and B.A. Sabel. 2000. Body distribution of 3H-labelled dalargin bound to poly(butyl cyanoacrylate. nanoparticles after i.v. injections to mice. Life Sciences **66**: 495-502.

Seed, J.R., and S.J. Black. 1997. A proposed density-dependent model of long slender to short stumpy transformation in the African trypanosomes. Journal of Parasitology **83**: 656-662.

----------, and S.J. Black. 1999. A revised arithmetic model of long slender to short stumpy transformation in the African trypanosomes. Journal of Parasitology **85**: 850-854.

----------, and J. Sechelski. 1988. Growth of pleomorphic *Trypanosoma brucei rhodesiense* in irradiated inbred mice. Journal of Parasitology **74**: 781-789.

-----------, and J.B. Sechelski. 1989. Mechanism of long slender (LS. to short stumpy (SS. transformation in the African trypanosomes. Journal of Protozoology **36**: 572-577.

Selzer, P.M., F. Hesse, B. Hamm-Kunzelmann, K. Muhlstadt, H. Echner, and M. Duszenko. 1996. Down regulation of S-adenosyl-L-methionine decarboxylase activity of *Trypanosoma brucei* during transition from long slender to short stumpy-like forms in axenic culture. Eur. Journal of Cell Biology **69**: 173-179.

Shapiro, S.Z., J. Naessens, B. Liesegang, S.K. Moloo, and J. Magondu. 1984. Analysis by flow cytometry of DNA synthesis during the life cycle of African trypanosomes. Acta Trop. **41**: 313–323.

Sperandio, S., I. de Bell, and D.E. Bredesen. 2000. An alternative, nonapoptotic form of programmed cell death. Proceedings of the National Academy of Sciences, USA **97**: 14376–14381.

Stuart, K., T.E. Allen, M.L. Kable, and S. Lawson. 1997. Kinetoplastid RNA editing: complexes and catalysts. Current Opinion in Chemistry and Biology **1**: 340-346.

Tasker, M., J. Wilson, M. Sarkar, E. Hendriks, and K. Matthews. 2000. A novel selection regime for differentiation defects demonstrates an essential role for the stumpy form in the life cycle of the African trypanosome. Molecular Biology of the Cell **11**: 1905-1917.

Taylor, E.M., D.A. Otero, W.A. Banks, and J.S. O'Brien. 2000. Designing stable blood-brain barrier-permeable prosaptide peptides for treatment of central nervous system neurodegeneration. Journal of Pharmacological and Experimental Therapy **293**: 403-409.

Turner, C.M.R., N. Aslam, and C. Dye. 1995. Replication, differentiation, growth and the virulence of *Trypanosoma brucei* infections. Parasitology **111**: 289-300.

Tyler, K.M., K.R. Matthews, and K. Gull. 1997. The bloodstream differentiation-division of *Trypanosoma brucei* studied using mitochondrial markers. Proceedings of the Royal Society of London B Biological Sciences **1387**: 1481-1490.

Van Den Abbeele, J., Y. Claes, D. van Bockstaele, D. Le Ray, and M. Coosemans. 1999. *Trypanosoma brucei* spp. development in the tsetse fly: characterization of the post-mesocyclic stages in the foregut and proboscis. Parasitology **118**: 469-478.

Vanhamme, L., and E. Pays. 1995. Control of gene expression in trypanosomes. Microbiology Reviews **59**: 223-240.

Vassella, E., and M. Boshart. 1996. High molecular mass agarose matrix supports growth of bloodstream forms of pleomorphic *Trypanosoma brucei* strains in axenic culture. Molecular and Biochemical Parasitology **82**: 91–105.

-----------, B. Reuner, B. Yutzy, and M. Boshart. 1997. Differentiation of African trypanosomes is controlled by a density sensing mechanism which signals cell cycle arrest via the cAMP pathway. Journal of Cell Science **110**: 2661-2671

Vickerman, K. 1965. Polymorphism and mitochondrial activity in sleeping sickness. Nature **208**: 762–766.

-----------. 1985. Developmental cycles and biology of pathogenic trypanosomes. British Medical Bulletin **41**: 105–114.

Welburn, S.C., and I. Maudlin. 1997. Control of *Trypanosoma brucei brucei* infections in tsetse, *Glossina morsitans*. Medical and Veterinary Entomology **11**: 286-289.

-----------, and N.B. Murphy. 1998. Prohibitin proto-oncogene and RACK homologues are up-regulated in trypanosomes induced to undergo apoptosis and in naturally occurring terminally differentiated forms. Cell Death and Differentiation **5**,:615-622.

WHO Expert Committee. 1998. Control and surveillance of African trypanosomiasis. World Health Organ. Tech. Rep. Ser. 881, I-VI, 1-114.

Wijers, D.J.B., and K.C. Willett. 1960. Factors that may influence the infection rate of *Glossina palpalis* with *Trypanosoma gambiense*: II. The number and morpholgy of the trypanosomes present in the blood of the host at the time of the infected feed. Annals of Tropical Medicine and Parasitology **54**: 341–350.

Yasaka, T., S. Ichisaka, T. Katsumoto, H. Maki, M. Saji, G. Kimura, and K. Ohno. 1996. Apoptosis involved in density-dependent regulation of rat fibroblastic 3Y1 cell culture. Cell Structure and Function **21**: 483-489.

Ziegelbauer, K., M. Quinten, H. Schwarz, T.W. Pearson, and P. Overath. 1990. Synchronous differentiation of *Trypanosoma brucei* from bloodstream to procyclic forms *in vitro.* European Journal of Biochemistry **192**: 373–378.

----------, Stahl, M. Karas, Y.D. Stierhof, and P. Overath. 1993. Proteolytic release of cell surface proteins during differentiation of *Trypanosoma brucei.* Biochemistry **32**: 3737–3742.

ENDOCYTOSIS IN AFRICAN TRYPANOSOMES

Derek .P. Nolan, Jose A. Garcia-Salcedo, Maurice Geuskens, Didier Salmon, Françoise Paturiaux-Hanocq, Annette Pays, Patricia Tebabi, and Etienne Pays

Laboratory of Molecular Parasitology, Institute of Molecular Biology and Medicine, 12 Rue des Profs. Jeener et Brachet, Gosselies, B-6041, Belgium

ABSTRACT

African trypanosomes face a peculiar dilemma in mammalian hosts. They are simultaneously constrained to avoid interactions that might be detrimental to their survival but also must acquire macromolecules. At the moment precisely how they have successfully resolved this conundrum remains elusive but what is clear is that they have developed a very extraordinary and efficacious machinery of endocytosis. Ligand binding and uptake occur only in a specialized region of the cellular surface called the flagellar pocket. Emerging evidence suggests that trypanosomal receptors for host macromolecules are unusual and that mechanisms of their internalization are different to that of higher eukaryotes. This review will focus on areas of endocytosis where there is general agreement, highlight conflicting views and consider general paradigms of the process.

Key Words Trypanosome; Flagellar Pocket; Endocytosis; Receptors

THE PARADOX OF ENDOCYTOSIS IN TRYPANOSOMES

African trypanosomes are extracellular parasites of mammals. Key to their biological success has been their ability to colonize the vasculature and interstitial spaces of their mammalian hosts. This achievement has necessitated the evolution of strategies to adapt to the mammalian immune response. The paramount mechanism here is antigenic variation, by which trypanosomes vary their surface coat. The coat consists of a single, tightly packed and highly immunogenic protein species (the variant surface glycoprotein or VSG) that covers the entire cellular surface. Variation of their coat glycoprotein, even at low frequency, always allows some trypanosomes to avoid immune destruction.

VSG may cloak the trypanosome surface, but it cannot mask all conserved antigens. Trypanosomes depend on their host for certain macromolecular growth factors, notably serum low and high density lipoproteins (Black and Vandeweerd, 1989) and transferrin (Black and Vandeweerd, 1989; Schell *et al.*, 1991b). These they take up by processes that are orders of magnitude more efficient than fluid phase endocytosis, and consequently involve receptors. Given the size of the ligands it is unlikely that their putative receptors are shielded by VSG. Moreover, functional constraints, *e.g.* the binding of specific ligands with high affinity, are likely to restrict significantly the degree to which receptors can imitate the VSG and undergo antigen variation of exposed domains. The requirement for macromolecular nutrients highlights a vexing paradox: how do the parasites take up these macromolecules without exposing the molecular machinery involved to the attention of the immune system?

A POCKET SIZED SOLUTION TO A CELLULAR PROBLEM

Endocytosis, exocytosis and possibly surface membrane turn over in general in bloodstream forms of African trypanosomes is restricted to a small invagination of the plasma membrane at the base of the flagellum called the flagellar pocket (FP) (reviewed in Webster and Russell, 1993) (Fig. 1). The surface membrane of the flagellum and FP lacks the closely spaced subpellicular array of microtubules associated with the cell body that are thought to be inimical to membrane vesicle fusion and fusion events essential for membrane traffic. Apart from the FP, the general outline of the endocytic pathway in trypanosomes is similar to that of higher eukaryotes. Invaginations of the FP membrane appear to pinch off to form coated vesicles that subsequently fuse with collecting tubular structures or multivesicular bodies located close to the FP (Fig. 2). These structures probably represent early endosomal compartments and appear to fuse with a larger, lysosomal-like digestive vacuole that appears to be the terminal point for cellular digestion. These compartments are acidic with the lysosomal compartment having an estimated pH of about 5.5 (Nolan and Voorheis, 2000).

At first glance it may seem that the limited area of the FP might act as a bottleneck for endocytosis but this is far from the case (Overath *et al,* 1997). All available evidence suggests that the FP is a site of tumultuous membrane traffic and out performs other eukaryotes in this respect by between one and two orders of

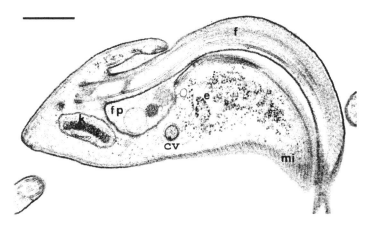

Figure 1. Longitudinal section of the posterior end of a bloodstream form of Trypanosoma brucei in the region of the base of the flagellum (f). The flagellar pocket (fp) is clearly visible and contains heterogeneous material and globular particles of debris. Also visible in the section are: the kinetoplast (k), endoplasmic reticulum (e), coated vesicles (cv) and the subpellicular array of microtubules (mi). Bar = 0,5 μM

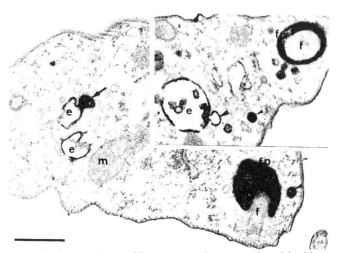

Figure 2. Bloodstream forms of Trypanosoma brucei incubated for 30 min with Tf-HRP, fixed and processed for cytochemical detection of HRP before embedding, sectioning and electron microscopy observation. The tagged ligand is concentrated in the flagellar pocket (fp). Coated vesicles (small arrow heads) formed by invagination of flagellar pocket membrane contain the ligand. These vesicles fuse with larger endosomes (e) and loose their coat. Small vesicles formed by budding and pinching off in the interior of the endosomes segregate to one pole to form multivesicular bodies (large arrow head) which detach then from the early endosomes. The intracellular trafficking pathway of the ligand includes tubular vesicular elements, and terminates in lysosome-like digestive vacuoles (Salmon et al. 1997).f: fagellum. m: mitochondria. Bar = 0,5 μM

magnitude. It has been calculated that the FP membrane is recycled about thirty times per hour. These estimates are based on clearance from the FP; by corollary they must be matched by a similar rate of membrane fusion events into the FP membrane. So this is a very busy traffic junction indeed! How such high rates of trafficking are sustained remains to be established and this is an issue of relevance to endocytosis in eukaryotic organisms in general. At least part of the solution is likely to be the sequestration of the proteins involved within the FP. Since the FP membrane is continuous with the plasma membrane this necessitates mechanisms that discriminate between proteins destined to remain in the FP and those intended to be uniformly distributed over the cellular surface. Resolving of this problem requires a molecular description of the endocytotic machinery of the FP. The place to start is with the best-characterized receptor in African trypanosomes: the transferrin receptor (Tf-R).

TRANSFERRIN UPTAKE BY TRYPANOSOMES.

Trypanosomes depend on host Tf for iron and uptake of Tf is specific, rapid, saturable and limited to the FP (Coppens *et al,* 1987). Work by several groups has established that the trypanosomal Tf-R is a heterodimer encoded by two genes originally identified by studies on antigenic variation. These studies demonstrated that VSG gene transcription occurs in telomeric units called expression sites (ESs). Although there are multiple ESs only one is active at a given time, and only in bloodstream forms. Equally interesting was the finding that ESs contained up to eleven associated genes (ESAGs) whose expression is linked with that of the VSG (reviewed in Pays *et al,* 2001). Most ESAGs possess no significant homology with other sequences and their function remains elusive.

The Tf-R is a 1:1 heterodimer composed of a heterogeneous glycoprotein of 50-60 kDa encoded by ESAG6 and a 42 kDa glycoprotein encoded by ESAG7 (Solman *et al.*, 1994). The Tf-R binds one molecule of Tf and modeling suggests that it forms a rod-shaped complex mounted normal to the plasma membrane that is about 70 % the axial length of a VSG dimer (Steverding, 2000). The entire complex is attached to the membrane by a GPI anchor located at the C-terminus of the ESAG 6 subunit and this anchor can be labeled with myristic acid (Nolan and Pays, unpublished), as is the case with the VSG GPI anchor. The Tf-R copy number is low, probably around 3000 receptors/cell, and it seems to have similar affinities for holo- and apo-Tf (Steverding, 1998), with apparent dissociation constants (K_d) in the nanomolar range (see Steverding, 2000). However, there is wide variability in receptor affinities among trypanosome strains for transferrin from different mammals with some receptors having miserable affinities indeed!

Once internalized the complex is routed via endosomes to lysosomes. Data obtained using pH sensitive probes suggest that ligand release from the receptor might occur early in the pathway, probably in the early endosome (H.P. Voorheis, personal communication). In any case the internalized Tf is rapidly degraded and the low molecular weight breakdown products are then released into the extracellular medium (Steverding *et al*, 1995). The fate of the released iron remains unknown but mechanisms must exist for its intracellular transport and probably storage or excretion because bloodstream trypanosomes require only 40,000 atoms per generation doubling time, which is eight times less than they take up (Steverding, 1998). In contrast, to endocytosed Tf, the Tf-R is recycled (Kabiri and Steverding, 2000) approximately 60 times at about 11 min per cycle before being degraded.

The heterodimeric Tf-R resembles a VSG dimer and each subunit folds like the *N*-terminal domain of the VSG (Salmon et al., 1997) with which it might share a common evolutionary origin. This study also showed that regions of major sequence difference between the different subunits form surface exposed loops on the rod-shaped complex and generate the ligand-binding site. Substitutions in these regions could increase as well as decrease the affinity of the receptor for a given host species of transferrin, consistent with the likelihood that polymorphism of Tf-Rs encoded in different ESs, may be a mechanism to overcome transferrin diversity between different mammal hosts.

TRANSFERRIN UPTAKE: WHAT REMAINS TO BE DONE?

Despite these advances the story of Tf uptake remains incomplete and key questions remain to be resolved. Most important is probably how is the GPI-anchored Tf-R constrained to the FP and internalised, and what role does extracellular Tf-R play in this process? Immunogold localization studies have consistently demonstrated that very little of the Tf-R is actually present at the surface of the FP and most is in the lumen of the FP and coated vesicles (Salmon *et al*, 1994; Steverding *et al*, 1995). Moreover, this luminal Tf-R was also present in trypanosomes lacking the GPI-specific phosphoinositol lipase C responsible for cleavage of the VSG anchor (Geuskens *et al*, 2000). Therefore, its generation may be mediated by another lipase. Alternatively, the nature of the lipid anchor may allow for equilibrium between membrane and luminal forms of the Tf-R. For example, the GPI-anchor might have only a single myristic acid chain and, thus, the Tf-R may be able to slip in and out of the membrane (Peitzsch and McLaughlin, 1993). Interestingly, such a situation also holds true for the products of a pair of ESAG7/6-related genes termed PAG1 and PAG2 (Koenig-Martin *et al*, 1992). Perhaps these heterodimers only interact weakly with the surface membrane, and for the most part detach from it. Irrespective of the origin of the luminal Tf-R, its potential role as an intermediate in Tf uptake deserves consideration since soluble forms of the receptor can bind Tf (Salmon *et al*, 1994). Whether Tf binding is required for receptor uptake is another open question but endocytosis of a luminal Tf-R obviously would require more than simple membrane flux from the FP.

TRANSFERRIN UPTAKE: TOWARDS A MODEL FOR ENDOCYTOSIS

The FP in bloodstream forms appears to be filled with a glycoproteinic, electron opaque material, notably absent in procyclic forms, that is also found in coated pits and vesicles (Langreth and Balber, 1975). Interestingly, the luminal Tf-R always appears to be associated with this material (Steverding *et al*, 1995). Perhaps the Tf-R partitions between the FP surface and associated luminal proteins, binds transferrin and is then internalized. However, this would obviously require a luminal mechanism to allow receptors to engage the endocytotic machinery, and either selective recovery of Tf-R-ligand complexes, or futile endocytosis and processing of unoccupied passenger receptors and matrix material.

Is there evidence for such a mechanism? Well, recent results are encouraging since it has been reported that N-glycans containing linear poly-N-acetyllactosamine (pNAL) might be involved in endocytosis in bloodstream trypanosomes (Nolan *et al*, 1999). These glycans are present only on luminal or extracellular domains of proteins, consist of linear repeats of $(Gal\beta 1 \rightarrow 4GlcNAc\beta 1 \rightarrow 3)_n$, and bind specifically to tomato lectin (TL). The pNAL moiety is implicated in trypanosome endocytosis in mammals on three counts: (i) it is present on glycoproteins of the flagellar pocket and endosomes of bloodstream stage and metacyclic trypanosomes, (ii) it is not present in any other sub-cellular location or trypanosome life cycle stage, and (iii) metacylic and bloodstrem form trypanosomes are the only differentiative stages able to endocytose Tf (Webster and Fish, 1989).

Although the Tf-R is composed of both ESAG6 and 7 subunits and both bind to TL under native conditions, only the

ESAG6 subunit is retained using lysates prepared by boiling in SDS, which suggests that the pNAL glycan is present only on ESAG6 (Nolan and Pays, unpublished). Significantly, chito-oligosaccharides that inhibit linear pNAL binding to TL specifically also substantially inhibit the uptake of Tf. This leads to the proposal that endocytosis of Tf, might involve interactions between pNAL glycans of receptors and a FP protein with TL-like specificity. We have now identified a bloodstream trypanosome protein (~ 45 kDa) that appears to bind to pNAL glycans (Nolan and Pays, unpublished data), but much remains to be done to establish whether or not glycans have such a role in endocytosis.

It may be that additional proteins are involved in the endocytic process. Indeed, the ESAG1 gene product may have a role. We have recently shown that ESAG1 is located in the FP and although N-glycosylated, it lacks pNAL chains (Nolan and Pays, unpublished). ESAG1 has three domains: a large extracellular/luminal domain containing N-linked carbohydrate, a single transmembrane span and short cytoplasmic domain (~ 30 residues). Uniquely among all other ESAGs and surface proteins described to date, this cytoplasmic domain contains two classical sets of vicinal leucine residues thought to act as recruitment signals for adaptor protein complexes and the assembly of coated pits/vesicles (Mellman, 1996). Significantly, peptides designed around these residues from ESAG1 appear to interact with a μ-like protein of adaptor protein complexes (Nolan, Michel and Pays unpublished). It is tempting to speculate that ESAG1 might provide the link between the internal endocytic machinery and a luminal lectin-like mechanism for recruitment and retention of the Tf-R and other proteins containing pNAL.

Overall, this mechanism has many attractions and would allow the cell to concentrate receptors in the FP while circumventing surface area restrictions. In terms of receptor-mediated endocytosis this may be more than an eclectic concern, since calculations have shown that the surface area of the FP membrane can only accommodate about 50,000 VSG-sized polypeptides (Overath *et al*, 1992). It is worth noting that while the pNAL chain associated with the Tf-R is probably short, other FP glycoproteins appear to have extensive pNAL chains, e.g. ISG100 and ESAG2 (Nolan *et al*, 1999). Possibly these glycans are spatially extended within the FP and might form part of the electron opaque matrix characteristic of bloodstream forms.

UPTAKE OF LIPOPROTEINS BY BLOODSTREAM TRYPANOSOMES

T. brucei has an absolute requirement for lipoproteins for growth and G1 progression which can be satisfied by low or high density lipoproteins (LDL, HDL) from any of several different mammal species (Black et al., 1989). Unlike iron derived from Tf uptake, it seems that trypanosomes have a continuous requirement for lipoproteins, arresting in G1 of the cell cycle and undergoing terminal differentiation in their absence (Morgan *et al,* 1993, 1996). This requirement makes the trypanosome and lipoprotein interaction an attractive target for immune inhibition. Indeed the huge size of lipoproteins, LDL has a diameter of 20 – 25 nm and HDL a diameter of 15 – 20 nm, determines that their receptors must be free of masking VSG. It is therefore ironic, that, compared to Tf, our understanding of how trypanosomes interact with lipoproteins remains at a rudimentary level.

It has been shown that *T. brucei* acquires lipids from LDL and HDL with orders of magnitude higher efficiency than can be accounted for by fluid phase endocytosis suggesting receptor-mediated uptake (Vandeweerd and Black, 1990a). However, the parasites cannot acquire lipids from the very large lipoproteins, namely, chylomicrons and very low density lipoproteins. This led to the idea that chylomicrons and VLDL, which share apolipoproteins with HDL, may not be able to access the lipoprotein receptor(s) on *T. brucei*, consistent with a FP location. Bloodstream stage *T. brucei* were shown to acquire lipoprotein-associated cholesterol, cholesteryl esters, and phospholipids by pathways that could each be selectively inhibited by agents added to the incubation medium, but without acquisition of associated apolipoproteins (Vandeweerd and Black 1990b). Furthermore, the lipid uptake did not require exogenous divalent ions and was insensitive to weak bases and chloroquine. It was suggested that the particles may be disassembled in the FP or an early, non-acidified or only weakly acidified endosome, lipids acquired and unused material returned to the medium via the FP. How then is this achieved?

An early study on endocytosis in trypanosomes found that LDL uptake occurred at a rate between two and three orders of magnitude greater than BSA and that ligand binding was specific and saturable (Coppens *et al,* 1987). These features were consistent with the presence of a specific receptor for LDL. A Scatchard analysis of LDL binding suggested that each trypanosome has 52,000 low-affinity binding sites ($K_d \sim 250$ nM) and about 1800 high-affinity binding sites with a K_d of 6 nM (Coppens *et al,* 1988). The same study also reported the isolation, using immobilized LDL particles as an affinity step, of a protein of 86 kDa that bound LDL. Antibodies against this protein labeled primarily the FP but also to a lesser extent the flagellar membrane outside of the pocket. These antibodies inhibited LDL binding at 4°C and growth of *T. brucei* in culture (Coppens *et al,* 1988). It was sugggested that the high affinity LDL-R might be located in the FP with the low affinity form being located along the flagellum. However, it is not clear that these low-affinity binding sites actually represent receptors at all (see Borst and Fairlamb, 1998). A later study indicated that the LDL receptor (LDL-R) was actually a protein of 145 kDa and that the 86 kDa protein represented a proteolytic product of this polypeptide (Coppens et al 1991). Additional studies went on to reveal that processing of the endocytosed LDL involved an acidic compartment and it was proposed that receptor recycling occurred due to acid-induced ligand-receptor dissociation (Coppens *et al,* 1993). The protein constituents of endocytosed LDL were degraded and the cell retained some of this material but most was released into the extracellular medium. Remarkably, the same LDL-affinity purification step allowed the isolation of a 145 kDa protein from variety of Kinetoplastida, including organisms that have never encountered a mammal, and antibodies against the *T. brucei* protein recognized this protein as well the mammalian receptor (Bastin *et al,* 1994; 1996). It was also apparently possible to isolate, by immunoprecipitation or an LDL-affinity step, a labeled 145 kDa protein from detergent lysates of *Crithidia* or procyclic forms of *T. brucei* metabolically labeled with [^{35}S] methionine/cysteine (Bastin *et al,* 1996).

Notwithstanding these results, doubts remain whether this 145 kDa protein is involved in acquisition of LDL by T. brucei for growth (reviewed in Black et al, 2001). In addition, it is certainly troubling that it has so far proved impossible to clone the corresponding gene even using a bank of monoclonal antibodies. Other reviewers, concerned about the presence of the same LDL-R in widely divergent organisms, have argued that the 145-kDa protein is not the LDL-R but might be an evolutionary conserved and relatively abundant protein that co-purifies during the affinity step (Borst and Fairlamb, 1998). Given that the overall purification factor was only 1000-fold this is a plausible argument. There is also the possibility that this protein may be derived from the LDL complex itself. Recall that the affinity purification step employed in these experiments used immobilized LDL particles and unless all protein constituents of the complex were fully and covalently attached to the resin the possibility of some protein leakage from the resin cannot be fully discounted. A key question concerns the reproducibility of the metabolic labelling experiments. Interestingly, we found that an antibody against the 145 kDa protein recognized faintly a band of similar size in Western blots of the TL binding fraction isolated from *T. brucei* (Nolan *et al*, 1999). We now think that this might be due to co purification of a small amount of LDL ligand actually bound to the receptor since the isolations were performed under native conditions and we suspect that the LDL-R like the Tf-R contains N-linked linked pNAL and binds to TL.

Trypanosomes also appear to have receptors for HDL. This evidence is largely from studies on the binding of labeled HDL to trypanosomes at 4 °C and there is some scatter in the data. Variable number of receptors of moderate affinity per cell have been reported: 64,000 receptors with a K_d of 157 nM for human HDL and 11,500 receptors with a K_d of 315 nM for bovine HDL (Gillet and Owen, 1992), while Hager et al (1994) reported the presence of 22,000 sites for human HDL with half maximal binding at 80 µg/ml (~ 140 nM). This numbers of receptors seem rather large to be accommodated in the FP. It is unclear whether different receptors are involved in LDL and HDL uptake but there appears to be distinct receptors for different subclasses of HDL particles (Hager *et al*, 1994; Hager and Hajduk, 1997).

UPTAKE OF LIPOPROTEINS BY PROCYLIC FORMS.

Until recently it seemed that endocytosis was less important in insect midgut or procyclic forms of the parasite (Langreth and Balber, 1995). Indeed the only forms from tsetse flies apparently able to take up Tf were mammalian infective metacyclic forms from the tsetse fly salivary glands (Webster and Fish, 1989). This view was subsequently reinforced by the finding that even expression of the Tf-R in procyclic forms did not allow endocytosis of Tf (Ligtenberg *et al*, 1994). Recent developments may now require some revision of this view since Lee *et al* (1999) have reported HDL uptake by culture procyclic forms, although uptake of LDL was not detected. Uptake of HDL was dependent on the growth phase of the parasites and varied to some extent with the quality of the ligand. Once internalized the HDL appeared to be degraded in acidic intracellular compartments. Subsequent studies revealed that CRAM, a highly reiterative, transmembrane protein located in the FP, might be a receptor for HDL in procyclic forms of *T. brucei* (Lee *et al*, 1990; Liu *et al*, 2000). This protein is far more abundant in procylic than bloodstream forms and has homologies with the cysteine-rich repetitive motif characteristic of the human LDL receptor, which was the basis of the original idea that this protein might be involved in LDL uptake. Significantly, binding of HDL in CRAM procyclic null mutants was reduced by about 80% but ligand uptake apparently decreased only by 20%, which suggests that CRAM is not solely responsible for HDL uptake (Liu *et al*, 2000). Analysis of CRAM deletion mutants indicated that the protein is routed to and retained in the FP by virtue of signals localized in the short C-terminal cytoplasmic domain of the protein, although this domain contains no obvious sorting motifs (Yang *et al*, 2000). Even though a direct demonstration of HDL binding to CRAM is lacking, these are nonetheless encouraging, and potentially groundbreaking, results. It would be more than a little ironic if studies on procyclic forms were to point the way forward to understanding lipoprotein uptake by trypanosomes given the past focus on bloodstream forms. An obvious question is whether CRAM also might be the receptor in bloodstream forms? This is probably unlikely since CRAM is ten fold less abundant in bloodstream forms but HDL binding seems to be the same as in procyclic cells.

A BRIEF VIEW OF THE INTERNAL MACHINERY

Little is known about the details of the internal endocytic machinery in trypanosomes. What is clear is that membrane traffic in trypanosomes, as in other eukaryotes, occurs by budding and fusion of vesicles. There is also evidence that sorting of internalized surface proteins into separate pathways occurs. For example, transferrin-gold particles and VSG molecules internalized in the same endocytic vesicles apparently enter into tubular-like endosomes and are then segregated into different structures (Webster and Grab, 1988). A similar segregation of VSG and gp65, a resident protein of endosomes, has been reported in *T. vivax* (Burleigh *et al*, 1993). Interestingly gp65 colocalized maximally with endocytosed BSA after 5 min at 37°C and hence might be a component of an early endosomal compartment. These results suggest that all internalized surface proteins are sorted into separate pathways of traffic. Morphological studies have also shown that the cytoplasmic face of the FP membrane is lined with a bristle or "spiny" coat and that invaginations of the FP membrane pinch off to form similarly coated endocytic vesicles (Webster, 1989). This coat is present in bloodstream and mammalian metacyclic forms but absent in procyclic forms. And hence its developmental regulation resembles that of linear pNAL glycans. Whether this coat contains clatherin or is clatherin-like is unknown but given the association with endocytic vesicles it seems a reasonable possibility. Significantly, the trypanosome genome project has already uncovered sequences that share significant homology with the mammalian clatherin heavy chain as well as other potential homologues of adaptor (AP) and coatomer (COP) protein complexes involved in endocytosis and membrane traffic in higher eukaryotes (Mellman 1996; Scales *et al*, 2000).

There is also good evidence for the involvement of other non-coat proteins related to those involved in membrane traffic in other eukaryotes. An exemplar here is the Rab family of small GTPases that play an essential role in membrane vesicle trafficking by imparting specificity to vesicle targeting. In trypanosomes multiple forms of these proteins have now been identified and termed TbRabs by the Field group, which has been most prominent in investigating their role in *T. brucei*. Different TbRabs appear to be involved in different pathways of membrane traffic. For example, TbRab5 is clearly a component of the endocytic pathway and co-localizes with ISG100, a resident protein of the pathway, while TbRab4 appears to be largely present in a different compartment to TbRab5 and might play a role in recycling (Field et al, 1998). Taken together, these considerations suggest that the internal machinery involved in endocytic/membrane traffic is likely to follow paradigms described in other eukaryotes.

WHAT REMAINS TO BE DISCOVERED?

Clearly, many furrows remain to be ploughed in this particular field. Hopefully, these will not be lonely paths since there is ample justification for the task ahead. Understanding brings knowledge and the power to act. If the ability of these parasites to take up essential macromolecules can be compromised, perhaps even only slightly, then potentially a major burden on Africa might be lifted. The laudable aim of this volume of reviews is to make this case as eloquently as possible because of the depressing but currently fashionable view, promulgated by some scientists and administrators of research budgets, that trypanosomiasis is an intractable problem and that basic research in this field serves only to satisfy academic curiosity. Nothing could be further from the truth. About a decade ago our understanding of the molecular basis of receptors and endocytosis in trypanosomes was in the doldrums. Discoveries since then, driven entirely by basic research on these processes, have transformed the situation and should act as a talisman for those of us who believe that such research efforts represent the way forward. If the exigencies of trypanosomiasis were not ample enough reason for a continued research effort on endocytosis in African trypanosomes the evidence in the literature provides additional justification. By any criterion, whether biochemical, molecular or cellular, trypanosomes have more than punched their weight in terms of a contribution to our understanding of eukaryotic biology. Almost a century after their discovery, studies on these unicellular eukaryotes continue to propel productive areas of research with wider implications for other organisms and there is no expectation that this usefulness will diminish in the near future.

Acknowledgements Work on endocytosis in the laboratory of molecular parasitology was supported by research contracts with the Communauté Française de Belgique and the Interuniversity Poles of Attraction Programme of the Belgian State Prime Minister's Office - the Federal Office for Scientific, Technical and Cultural Affairs. Author for correspondance: DPN, current address Dept. Biochemistry, Trinity college, Dublin 2, Dublin, Ireland.

REFERENCES

Bastin, P., A. Stephan, J.Raper, J.-M. Saint-Remy, F.R. Opperdoes, and P. Courtoy. 1996. An Mr 145 kDa low-density lipoprotein (LDL)-binding protein is conserved throughout the Kinetoplastida order. Mol. Biochem. Parasitol. **76**: 43-56.

------------, I. Coppens, J.M. Saint-Remy, P.Baudhuin, F.R. Opperdoes, and P.J. Courtoy. 1994. Identification of a specific epitope on the extracellular domain of the LDL-receptor of *Trypanosoma brucei brucei*. Mol. Biochem. Parasitol. **63**: 193-202.

Black, S.J., and V. Vandeweerd. 1989. Serum lipoproteins are required for multiplication of *T. brucei* under axenic culture conditions. Mo. Biochem. Parasitol. 37: 65-72.

-------------, S.J., J.R. Seed, and N.B. Murphy, 2001. Innate and acquired resistance to African trypanosomiasis. J. Parasitol. 87:1-9.

Borst, P., and A.H. Fairlamb. 1998. Surface receptors and transporters of *Trypanosoma brucei.* Annu. Rev. Microbiol. **52**: 745-778.

Burleigh, B.A., C.W. Wells, M.W. Clarke, and P.R. Gardiner. 1993. An integral membrane glycoprotein associated with endocytic compartment of *Trypanosoma vivax:* identification and partial characterization. J. Cell Biol. **120**: 339-352.

Coppens, I., F.R. Opperdoes, P.J. Courtoy, and P. Baudhuin. 1987. Receptor-mediated endocytosis in the bloodstream form of *Trypanosoma brucei.* J. Protozool. **43**: 465-473.

--------------, I., P. Baudhuin, F.R. Opperdoes, and P.J. Courtoy. 1988 Receptors for the low density lioproteins on the hemoflagellate *Trypanosoma brucei:* purification and involvement in the growth of the parasite. Proc. Natl. Acad. Sci. USA **85**: 6753-6757.

--------------, I., P. Bastin, P.J. Courtoy, P. Baudhuin, and F.R. Opperdoes. 1991. A rapid method purifies a glycoprotein of Mr 145,000 as the LDL receptor of *Trypanosoma brucei brucei.* Biochem. Biophys. Res. Commun. **178**: 185-191.

---------------, I., P. Baudhuin, F.R. Opperdoes, and P.J. Courtoy. 1993. Role of acidic compartments in *Trypanosoma brucei,* with special reference to low-density lipoprotein processing. Mol. Biochem. Parasitol. **58**: 223-232.

Field, H., M. Farjah, A. Pal, K. Gull, and Field. 1998. Complexity of trypanosomatid endocytosis pathways revealed by Rab4 and Rab5 isoforms in *Trypanosoma brucei*. J. Biol. Chem. **273**: 32102-32110

Geuskens M, E., Pays, and M.L.Cardoso de Almeida. 2000. The lumen of the flagellar pocket of *Trypanosoma brucei* contains both intact and phospholipase C-cleaved GPI anchored proteins. Mol Biochem Parasitol. **108**: 269-75

Gillet, M.P.T., and J.S. Owen. 1992. Characteristics of the binding of human and bovine high density lipoproteins by bloodstream forms of the African trypanosome , *Trypanosoma brucei brucei*. Biochim. Biophys. Acta **1123**: 239-248

Hager, K.M, M.A. PierceD.R. Moore, E.M.Tytler, J.D. Esko, and S.L. Hajduk. 1994. Endocytosis of a cytotoxic human high density lipoprotein results in disruption of acidic intracellular vesicles and subsequent killing of African trypanosomes. J. Cell Biol. **126**: 155-167.

---------------. and S.L. Hajduk. 1997. Mechanism of resistance of African trypanosomes to cytotoxic human HDL. Nature **385**: 823-826.

Kabiri, M., and D. Steverding. 2000. Studies on the recycling of the transferrin receptor in *Trypanosoma brucei* using an inducible gene expression system. Eur. J. Biochem. **267**: 3309-3314.

Koenig-Martin, E., M. Yamage, and I. Roditi. 1992. A procyclin-associated gene in *Trypanosoma brucei* encodes a polypeptide related to ESAG 6 and 7 proteins. Mol. Biochem. Parasitol. **55**: 135-145.

Langreth, S.M., and A.E. Balber. 1975. Protein uptake and digestion in bloodstream and culture forms of *Trypanosoma brucei*. J. Protozool. **22**: 40-53.

Lee, M.G.S., B.E. Bihain, D.G. Russell, R.J. Deckelbaum, and L.H.T. Van der Ploeg. 1990. Characterization of a cDNA encoding a cystein-rich cell surface protein located in the flagellar pocket of the protozoan *Trypanosoma brucei*. Mol. Cell. Biol. **10**: 4506-4517.

---------------, F.T. Yen, Y. Zhan, and Bihain, B.E. 1999. Acquisition of lipoproteins in the procycclic form of *Trypanosoma brucei*. Mol. Biochem. Parasitol. **100**: 153-162

Ligtenberg M.J., W. Bitter, R. Kieft, D. Steverding, H. Janssen, J. Calafat, and P. Borst. 1994. Reconstitution of a surface transferrin binding complex in insect form *Trypanosoma brucei*. EMBO J. **13**: 2565-73.

Liu, J., X. Qiao, D. Du, and M. G.-S. Lee. 2000. Receptor-mediated endocytosis in the procyclic form of *Trypanosoma brucei*. J. Biol. Chem. **275**:12032-12040.

Mellman, I. 1996. Endocytosis and molecular sorting. Ann. Rev. Cell Dev. Biol. **12**: 575-625

Morgan, G.A., H.B. Laufman, F.P. Otieno-Omondi, and S.J. Black. 1993. Control of G1 to S cell cycle progression of *Trypanosoma brucei* S427 organisms under axenic culture conditions. Mol. Biochem. Parasitol. **57**: 241-252.

-----------------, E.A. Hamilton, and S.J. Black 1996. The requirements for G1 checkpoint progression of *Trypanosoma brucei* S 427 clone 1. Mol. Biochem. Parasitol. **78**: 195-207.

Nolan, D.P., M. Geuskens, and E. Pays. 1999. N-linked glycans containing linear poly-N-acetyllactosamine as sorting signals in endocytosis in *Trypanosoma brucei*. Curr. Biol. **9**: 1169-1172.

--------------, and H.P. Voorheis. 2000. Hydrogen ion gradients across the mitochondrial, endosomal and plasma membranes in bloodstream forms of *Trypanosoma brucei*. Solving the three compartment. Eur. J. Biochem. **267**: 4601-4614.

Overath, P., D. Schell, Y.-D. Steirhof, H. Schwartz, and D. Preis, D. 1992. A transferrin-binding protein in *Trypanosoma brucei*: Does it function in iron uptake. In

Dynamics of Membrane Assembly, NATO ASI Ser., JAF Op den Kamp (ed.) Springer-Verlag, Berlin, H63 p. 333-347.

--------------,Y.-D. Steirhof, and M. Weise. 1997. Endocytosis and secretion in trypanosomatid parasites-tumultuous traffic in a pocket. Trends Cell Biol. **7**: 27-33.

Pays, E., S. Lips, D. Nolan, L. Vanhamme, D. Pérez-Morga. 2001. The VSG expression sites of *Trypanosoma brucei*: multipurpose tools for the adaptation of the parasite to various mammalian hosts. Mol. Biochem. Parasitol. In press.

Peitzsch, R.M, and S. McLaughlin. 1993. Binding of acylated peptides and fatty acids to phospholipid vesicles: Pertinence to myristoylated proteins. Biochemistry **32**: 10436-10443.

Salmon, D., M. Geuskens, F. Hanocq, J. Hanocq-Quertier, D. Nolan, L. Ruben, and E. Pays. 1994. A novel heterodimeric transferrin receptor encoded by a pair of VSG expression site-associated genes in *Trypanosoma brucei*. Cell **78**: 75-86.

--------------, J. Hanocq-Quertier, F. Parriaux-Hanocq, A. Pays, P. Tebabi, D.P. Nolan, A. Michel,. and E. Pays. 1997. Characterization of the ligand-binding site of the transferrin receptor of *Trypanosoma brucei* demonstrates a structural relationship with the N-terminal domain of the variant surface glycoprotein. EMBO J. **16**: 7272-7278.

Scales, S.J., M. Gomez, and T.E. Kreis. 2000. Coat proteins regulating membrane traffic. Int. Rev. Cytol. **195**: 67-144.

Schell, D., N.K. Borowy, and P. Overath. 1991b. Transferrin is a growth factor for the bloodstream form of *Trypanosoma brucei*. Parasitol. Res. **77**: 558-560

Steverding D., Y.-D. Stierhof, H. Fuchs, R. Tauber, and P. Overath. 1995. Transferrin-binding protein complex is the receptor for transferrin uptake in *Trypanosoma brucei*. J. Cell Biol. **131**: 1173-82.

----------------. 1998. Bloodstream forms of *Trypanosoma brucei* require only small amounts of iron for growth. Parasitol. Res. **84**: 59-62.

----------------. 2000. The transferrin receptor of *Trypanosoma brucei.* Parasitol. Int. **48**: 191-198.

Vandeweerd, V. and S.J. Black. 1989. Serum lipoprotein and *Trypanosoma brucei brucei* interactions *in vitro*. Mol. Biochem. Parasitol. **37**: 201-211.

--------------------. and S.J. Black. 1990. Selective inhibition of the uptake by bloodstream form *Trypanosoma brucei brucei* of serum lipoprotein-associated phospholipid and cholesteryl ester. Mol. Biochem. Parasitol. **41**: 197-206.

Webster, P. 1989. Endocytosis by African trypanosomes. I. Three-dimensional structure of the endocytic organelles in Trypanosoma brucei and T. congolense. Eur. J. Cell Biol. **49**: 295-302

--------------, and D.J. Grab. 1988. Intrcellular colocalization of variant surface glycoprotein and transferrin-gold in *Trypanosoma brucei.*J. Cell Biol. **106**: 279-288.

--------------, and W. R. Fish. 1989. Endocytosis by African trypanosomes. II. Occurrence in different life-cycle stages and intracellular sorting. Eur. J. Cell Biol. **49**: 303-310

--------------, and D.G. Russell. 1993. The flagellar pocket of trypanosomatids. Parasitology Today **9**: 201-206.

Yang, H., D.G. Russell, B. Zheng, M. Eiki, and M.G. Lee. 2000. Sequence requirements for trafficking of the CRAM transmembrane protein to the flagellar pocket of african trypanosomes. Mol. Cell Biol. **20**: 5149-5163.

THE GENOME OF THE AFRICAN TRYPANOSOME

John E. Donelson
Department of Biochemistry, University of Iowa, Iowa City, Iowa 52242, USA

ABSTRACT

Trypanosoma brucei is a diploid organism with a nuclear haploid DNA content of 35 ± 9 megabase pairs (Mb) depending on the trypanosome isolate. About 15% of the total cellular DNA is in the kinetoplast where it is organized as homogenous 23-kb maxicircles and heterogeneous 1-kb minicircles. The remaining 85% of the DNA occurs in the nucleus as linear DNA molecules ranging in size from 50 kb to 6 Mb. At least 11 pairs of megabase chromosomes of 1 to 6 Mb exist that are numbered I-XI from smallest to largest. The two homologues of a megabase chromosome pair can differ in size by as much as 4-fold. Several intermediate-sized chromosomes of 0.2 - 0.9 Mb and uncertain ploidy are also present. The telomeres of the megabase and intermediate chromosomes are linked to expression sites for the genes encoding the variant surface glycoproteins (VSGs). In addition, about 100 linear minichromosomes of 50-150 kb occur and serve as repositories for unexpressed, telomere-linked VSG genes. About 50% of the nuclear genome is coding sequence. To date only one tRNA gene and one protein-encoding gene, specifying poly(A) polymerase, have been found to contain an intron. The complete sequence determination of chromosomes I and II (about 1 Mb each) is nearing completion and more than 20 Mb of discontinuous single-pass genomic DNA sequence data have been generated. Based on analogy with the *Leishmania* genome, much of the African trypanosome nuclear genome is likely to be arrayed as long transcription units of 50 or more intronless genes. Knowledge of this genomic sequence and its complete set of genes will open many new avenues for identifying better ways to control or eliminate this pathogen and its deadly disease.

Key words trypanosome, karotype, genome, sequence, genes, antigenic variation, kinetoplast, nucleus

Several species of African trypanosomes exist, but this essay will focus entirely on the genome of *Trypanosoma brucei brucei* (*T. brucei*) because it is the most extensively investigated of the African trypanosome genomes and its complete sequence is currently being determined. Much of the interest in the DNA molecules of African trypanosomes has been due to (i) the unusual properties of the DNA in the kinetoplast and (ii) the phenomenon of antigenic variation used by the parasites to evade the immune response of its host. Each of these features will be described briefly, followed by a discussion of the chromosomal DNA in the *T. brucei* nucleus.

THE KINETOPLAST DNA OF *T. BRUCEI*

The first indication that the kinetoplast at the base of the flagellum of trypanosomes contains DNA came in 1898 with the report that both the kinetoplast and the nucleus stained bright red with the newly discovered Romanovsky counterstains (Ziemann 1898; Englund et al. 1995). This observation was reinforced in 1924 when it was also observed using the new Feulgen reaction that the kinetoplast has a large amount of DNA (Bresslau and Scremin 1924; Vickerman 1997). It is now known that the kinetoplast DNA (kDNA) is a giant network of interlinked supercoiled maxicircle and minicircle DNA molecules, constituting as much as 15% of the total DNA in the organism. The *T. brucei* kDNA network contains several dozen homogeneous maxicircle DNA molecules of 23 kb (Sloof et al. 1992) and several thousand heterogeneous minicircle DNA molecules of 1 kb (Chen and Donelson 1980). The maxicircle DNAs are equivalent to mitochondrial DNA in other eukaryotes. They encode ribosomal RNAs and about a dozen proteins, most of which are subunits of the multi-subunit mitochrondrial membrane complexes involved in oxidative phosphorylation. Remarkably, most of the primary maxicircle transcripts must undergo insertions or deletions of uridines at specific sites to generate an open reading frame and a functional mRNA, a process called RNA editing (for a review, see Stuart et al. 1997). The minicircles encode guide RNAs (gRNAs) of about 60 nucleotides each that serve as the templates for the RNA editing process. Each minicircle is entwined with an average of three neighboring minicircles and can encode as many as three gRNAs. The heterogeneous *T. brucei* minicircle population has about 300 different sequence classes, as measured by DNA renaturation kinetics (Donelson et al. 1979), indicating the potential for encoding as many as 900 different gRNAs.

The unique three-dimensional structure of the kDNA network offers equally unique topological challenges to its replication. Analyses of the kDNA before and after genetic crosses of *T. brucei* in the tsetse fly have shown that the maxicircles are primarily inherited uniparentally, whereas the minicircles are inherited from both parents (Turner et al. 1995; Gibson et al. 1997). During their replication the minicircles are first released individually from the network by a topoisomerase and then, after replication, their progeny reattach at two antipodal positions on the network periphery (Guilbride and Englund 1998). These distinctive kinetoplast features of RNA editing and kDNA network replication offer potential opportunities for the development of new trypanocidal drugs targeted against these processes that might not affect the corresponding processes of mRNA maturation and DNA synthesis in the mammalian host.

ANTIGENIC VARIATION IN *T. BRUCEI*

Antigenic variation in *T. brucei* is the phenomenon whereby the bloodstream trypanosomes periodically switch from one variant surface glycoprotein (VSG) on their surface to another in an effort to keep "one step ahead" of their mammalian host's immune response (Turner and Berry 1989; Turner 1997). The *T. brucei* genome contains about 1000 different VSG genes (*VSGs*), only one of which is expressed at a time with a few special exceptions (Van der Ploeg et al 1982; Esser and Schoenbechler 1985; Baltz et al. 1986; Muñoz-Jordán et al 1996). The unexpressed *VSGs* occur throughout the different chromosomes, including the minichromosomes (see below), but all expressed *VSGs* studied to date are located near the telomeres (for reviews see Borst et al. 1998; Cross et al. 1998; Donelson et al. 1998; Pays and Nolan 1998). The telomere-linked *VSG* expression sites (ESs) extend from a promoter to the telomeric repeats of $(TTAGGG)_n$ located downstream of the *VSGs* (Pays et al. 1989; Pedram and Donelson 1999). The conventional model of a bloodstream *VSG* ES (B-ES) is based primarily on the sequence of the B-ES for the AnTat 1.3 *VSG* (reviewed by Pays and Nolan 1998). This B-ES and the few others examined contain a polycistronic transcription unit spanning 45-60 kb. Downstream of the promoter are at least eight ES-associated genes (*ESAGs*), several kb of a 76-bp repeat and the *VSG*, followed by sub-telomeric and telomeric repeats. About 20 such B-ESs occur in the *T. brucei* genome (Zomerdijk at al. 1990; Navarro and Cross 1996), and a similar number of ESs exist for *VSGs* expressed during the metacyclic stage (M-ESs) (Lenardo et al. 1986; Turner et al. 1998), making a total of about 40 ESs. The telomere-linked M-BSs are organized as short (3-5 kb) monocistronic transcription units and generally lack all or most of the 76-bp repeats and *ESAG*s of B-ESs (Barry et al. 1998; Pedram and Donelson 1999; Alarcon et al. 1999).

The best-studied and most common *VSG* switching mechanism is a gene conversion, or "duplicative transposition", in which the *VSG* in an active B-ES is replaced with a duplicated copy of an unexpressed *VSG*. Other switching mechanisms include the duplicative conversion of an entire telomere plus its adjacent *VSG* to another chromosomal end (telomere conversion), and reciprocal exchange of two telomeres and their associated *VSG*s (telomere exchange). In addition, transcription of one B-ES can switch to another B-ES *in situ* without an associated DNA rearrangement (*in situ* activation). The M-ESs, on the other hand, are only activated by the *in situ* mechanism. Despite considerable effort (Rudenko et al. 1995; Horn and Cross 1997; Cross et al. 1998; Vanhamme and Pays 1998), the molecular events that activate transcription at one telomere-linked ES and silence expression at the other 40 B-ESs and M-ESs are not understood. A modified base, β-D-glucosylhydroxymethyluracil (base J), occurs in the silent ESs and is absent from the expressed ES, but this modification also occurs at internal repetitive sequences and is more likely to stabilize repression of silent ESs than to induce ES activation (van Leeuwen et al. 1998, 2000). Instead, *VSG* ES expression appears to be controlled by DNA sequence-independent, epigenetic mechanisms that probably involve the chromatin remodeling near the telomeres (Horn and Cross, 1997; Ersfeld et al. 1999). It will be a serious challenge to determine the nature of these three-dimensional regulatory changes, but a knowledge of the full genome sequence should help in the analysis.

THE NUCLEAR CHROMOSOMAL KAROTYPES OF *T. BRUCEI*

The nuclear chromosomes of *T. brucei* do not condense during mitosis, so it is not possible to use cytological methods to identify the number of chromosomes in the nucleus. However, *T. brucei* was among the first organisms whose DNA molecules were separated by pulsed field gel electrophoresis (PFGE) (Van der Ploeg et al. 1984), a technique used to resolve DNA molecules in the size range of 100 kb to several Mb. The use of PFGE, as well as DNA renaturation and cytophotometry, all indicate that the haploid nuclear

DNA content of *T. brucei* is about 35 Mb, with as much as 25% variation among isolates (Borst et al. 1982; Van der Ploeg et al. 1989: Gibson et al. 1992; Kanmogne et al. 1997; Hope et al. 1999). For comparison, the *Saccharomyces cerevisiae* genome contains 12 Mb and 6000 genes (Mewes et al. 1998), and the *Caenorhabditis elegans* genome has 97 Mb and 19,000 genes (*C. elegans* Sequencing Consortium 1998). Thus, the 35-Mb *T. brucei* nuclear genome probably harbors about 12,000 genes. A survey of the *T. brucei* genome by sequencing the ends of randomly sheared fragments (see below) suggests that about 50% of the genome is coding sequence (El-Sayed and Donelson 1997). Only one of the many *T. brucei* protein-coding genes examined to date has been found to possess an intron, *i.e.*, the poly(A) polymerase gene, which has a single 653-nucleotide intron (Mair et al. 2000). The only other gene known to have an intron is a tRNA gene, which has an 11-nucleotide intron (Schneider et al. 1994), although at least a few additional intron-bearing genes are likely to be found. The sequence of a small 269-kb chromosome of *Leishmania major* has been determined and found to encode two apparent transcription units: one of 50 tandem genes on one strand and one of 29 tandem genes on the other strand (Myler et al. 1999; McDonagh et al. 2000). By analogy, the organization of many, if not all, of the *T. brucei* genes is likely to be similar.

The nuclear chromosomes of *T. brucei* have been grouped into three general size classes based on their migrations in PFGE: the megabase chromosomes (1-6 Mb), the intermediate chromosomes (200-900 kb) and the minichromosomes (50-150 kb) (El-Sayed et al., 2000). The megabase chromosomes are diploid in the nucleus (Hope et al. 1999; Tait et al. 1989; Melville et al. 1998), whereas the intermediate chromosomes and minichromosomes are of uncertain ploidy. However, polyploidy of the megabase chromosomes can occur during both *in vitro* culture and *in vivo* transmission. In one experimental case, *i.e.*, when *T. brucei* was maintained in culture with mycophenolic acid, an inhibitor of IMP dehydrogenase, the number of copies of an entire 6-Mb chromosome increased 10-fold, an amplification that increased the total DNA content of the organism by 40% (Wilson et al. 1994). In other cases, hybrid progeny derived from experimental crosses of two *T. brucei* clones transmitted through flies were found to have a DNA content and Southern blot hybridization pattern consistent with the progeny being triploid rather than diploid (Wells et al. 1987; Gibson et al. 1992), although triploidy appears to be a relatively rare occurrence (Hope et al. 1999).

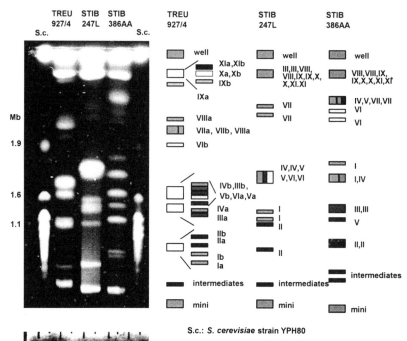

Fig. 1. *Molecular karotypes of three T. brucei field isolates, TRUE927/4, STIB257/L and STIB 386AA. The left panel shows an ethidium bromide-stained gel of the PFGE-separated chromosomes of the three T. brucei field isolates and of Saccharomyces cerevisiae (left and right lanes). The sizes of some of the S. cerevisiae chromosomes are indicated on the left. The right panel summarises the results of multiple Southern blot hybridizations of similar PFGEs with numerous cDNA probes used to identify the homologues of chromosomes I-VIII. The PFGE was conducted under conditions that are optimal for separating DNA molecules between 1 and 4 Mb. Other electrophoretic conditions are necessary to maximally resolve DNA molecules more than 4 Mb (chromosomes IX - XI) and less than 1 Mb [the intermediate chromosomes (intermed) and minichromosomes (mini)]. In this PFGE the DNA molecules larger than 4 Mb are in the compression zone or the gel well. Note that in TREU927/4 the chromosome III homologues are 1.7-1.8 Mb, whereas in STIB 247/L they are 4 Mb or larger (adapted from Melville et al. 1998; El-Sayed et al. 2000; and web site http://parsun1.path.cam.ac.uk/xsom.gif).*

The megabase chromosomes

   T. brucei contains at least 11 pairs of megabase chromosomes, which are numbered I to XI from smallest (~ 1 Mb) to largest (~ 6 Mb) as they occur in *T. brucei* stock TREU927/4 (Turner et al. 1997; Melville et al. 1998; El-Sayed, 2000). Letters are used to distinguish the two homologues, for example, chromosome Ia and Ib. Fig. 1, left panel, shows an ethidium bromide stain of a representative PFGE of three *T. brucei* stocks, including TREU927/4, conducted under electrophoretic conditions that maximize the separation of DNA molecules between 1 and 4 Mb. Under these conditions chromosomes larger than 4 Mb migrate in the compression zone or remain in the gel well, and the intermediate and minichromosomes are poorly resolved at the bottom of the gel. As summarized in the right panel of Fig. 1, Southern blots of similar PFGEs have been probed with more than 500 cDNAs to map the locations of genes on the chromosomes and to identify the chromosomal homologues (Melville et al. 1998, 1999). Many cDNAs hybridize to multiple chromosome pairs, but 10 or more different cDNAs have been identified that hybridize uniquely to each of the 11 chromosome pairs. These hybridizations demonstrate that synteny is usually maintained across *T. brucei* stocks (for a possible exception see Gibson and Garside 1991). However, the sizes of both the chromosomal pairs and the homologues within a pair vary greatly among *T. brucei* stocks (Melville 1997). For example, in TREU927/4 the chromosome III homologues are about 1.7 Mb, whereas in STIB247L they migrate in the compression zone and are larger than 4 Mb. Likewise, in STIB386AA the homologues of chromosomes I and II are about the same size but the chromosome IV homologues differ by at least 1 Mb. The causes of these enormous size polymorphisms are not well understood and can only be determined unambiguously by genome sequence determinations, but they are at least partially due to (i) rearrangements of *VSG*s and ESs, (ii) expansion/contraction of retrotransposon-like INGI and RIME sequences regions (Hasan et al. 1984; Murphy et al. 1987; Kimmel et al. 1987), tandemly arrayed housekeeping genes, subtelomeric/telomeric repeats and other repetitive sequences, and probably (iii) other unknown sequence changes (Melville et al. 1999).

A map of the 1.15-Mb chromosome Ia has been constructed using restriction digests and overlapping genomic DNA clones (Melville et al. 1999). As summarized in the top diagram of Fig. 2, hybridizations with various probes suggest that a B-ES is adjacent to one telomere of the chromosome and a possible M-ES is located next to the other. On the chromosomal interior side of the B-ES is a large region to which probes for the repetitive retrotransposon-like INGI and RIME sequences hybridize. This INGI/RIME region differs in size in the other chromosome I homologue and is responsible for most of the size polymorphism between chromosomes Ia and Ib in TREU927/4. The remaining interior portion of chromosome I contains primarily housekeeping and trypanosome-specific genes.

All indications are that the other megabase chromosomes are organized in a similar manner, *i.e.*, the B-ESs and M-ESs are adjacent to the telomeres and other genes, including unexpressed *VSGs*, are at interior sites. Interestingly, the diploid megabase chromosomes appear to harbor haploid ESs at their telomeres, *i.e.*, the ESs at the ends of chromosome Ia are not identical to the ESs at the ends of chromosome Ib. Hence, the approximately 40 telomere-linked ESs could be occur on the 11 megabase chromosome pairs, which have a total of 44 telomeres. Genome sequence determination should confirm the presence of these ESs and identify the boundaries separating the homologous and non-homologous segments of the chromosomal pairs.

Chromosome Ia (1.15 Mb)

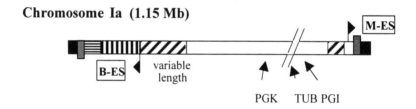

Minichromosomes (50 - 150 kb)

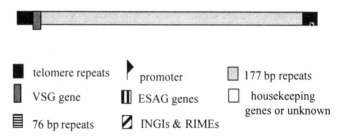

Fig. 2. *Diagrams of the 1.15-Mb chromosome Ia of T. brucei isolate 927/4 (top diagram) (Melville et al. 1997; Melville et al. 1999) and the 50-150 kb minichromosomes (bottom diagram) (Weiden et al. 1991) with an emphasis on the sequence elements near the telomeres. The telomere-linked expression sites for bloodstream VSGs (B-ES) are 45-60 kb in length, whereas telomere-linked expression sites for metacyclic VSGs (M-ES) are 3-5 kb in length. Repetitive INGI and RIME elements in chromosome I occur predominantly in a repetitive region of variable length on the chromosome-interior side of the B-ES. The housekeeping genes coding for phosphoglycerate kinase (PGK), tubulin (TUB) and phosphoglycerate isomerase (PGI) are unique to chromosome I. Minichromosomes are distinguished by their short length, a 177-bp repeat that comprises as much as 90% of their sequence and other minichromosome-specific repeats. The lengths of the different sequence elements are not drawn to scale (adapted from El-Sayed et al. 2000).*

The minichromosomes

The bottom diagram of Fig. 2 shows the general structure of a minichromosome that has been examined in detail (Sloof et al. 1983; Weiden et al. 1991). These linear minichromosomal DNA molecules of 50 – 150 kb possess the same $(TTAGGG)_n$ telomere repeats as the other larger *T. brucei* chromosomes. They are comprised predominately of internal tandem arrays of a 177-bp repeat, which constitute >90% of the sequence in some minichromosomes. Other minichromosome-specific repeat sequences that are either GC-rich or AT-rich occur between the telomeres and the interior 177-bp repeats. Some minichromosome telomeres are linked to silent *VSG*s, whereas others are not. To date none of the minichromosomes have been found to possess an active *VSG* ES. Therefore, to be expressed, these minichromosomal *VSG*s must undergo either an interchromosomal duplication or be part of a telomere exchange. Two minichromosomes of 55 and 60 kb have been found to have a rRNA gene promoter near one telomere, perhaps because of a recombination with a repetitive rRNA gene locus in a megabase chromosome (Zomerdijk et al. 1992). When a drug resistance gene, NEO^R, with the proper RNA processing signals was inserted downstream of this promoter in the 55-kb molecule, the resulting trypanosomes became drug resistant, and remained resistant through 130 generations without drug selection, indicating that the minichromosome is stably inherited during mitosis. It was not possible to determine whether these 55-kb and 60-kb DNA molecules are homologues, or to determine the ploidy of minichromosomes in general, because most of their sequence is the 177-bp repeat. Likewise, the reason for the existence of minichromosomes is not known, but if their only function is to serve as repositories for silent, telomere-linked *VSG*s, then a considerable portion of the nuclear genome, *i.e.*, 10-20%, is devoted to this purpose (Donelson 1996).

The intermediate chromosomes

These DNA molecules of 0.2 - 0.9 Mb in most *T. brucei* stocks are even more mysterious than minichromosomes. Their numbers and sizes vary among stocks (see Fig. 1) and they contain few, if any, unique markers or housekeeping genes. Their ploidy is uncertain and it is not known if they contain their own unique repeat sequences. They typically do not hybridize to the minichromosome-specific 177-bp repeat, although they can possess telomere-linked

*VSG*s or *VSG*-like sequences (Rudenko et al. 1998; Lips et al. 1993). It is possible they also serve as repositories for telomere-linked *VSG*s or even *VSG* ESs, although, similar to minichromosomes, it seems unnecessary for the trypanosome to devote this much DNA to solely that function.

In summary, the 140 or more DNA molecules in the *T. brucei* nucleus include at least 11 pairs of megabase chromosomes, and an indeterminate number of minichromosomes and intermediate chromosomes of uncertain function and ploidy. Unexpressed *VSG*s are located on most, if not all, of these chromosomes and likely consume 5% or more of the genome (Donelson 1996). The minichromosome-specific 177-bp repeats could constitute another 10-20% of the genome and the ~400 copies of the 5.2-kb INGI/RIME sequence probably accounts for another 5%, leaving about 70% of the nuclear genome for other genes and intergenic regions.

SEQUENCING THE GENOME OF *T. BRUCEI*

An African trypanosome genome network, sponsored by the UNDP/World Bank/WHO Special Programme for Research and Training in Tropical Disease (TDR), was initiated in 1995 to stimulate further characterization of the *T. brucei* genome and to generate the reagents for its sequence determination. The network is a loose association of a number of laboratories worldwide that conduct laboratory-based research on African trypanosomes. Activities of the network are described at the following Internet sites: http://www.ma.ucla.edu/par/ and http://parsun1.path.cam.ac.uk/.

Expressed Sequence Tags (EST) of *T. brucei*.

Single pass sequence determinations at the 5' ends of about 4500 cDNAs (ESTs) from *T. brucei* have been conducted (El-Sayed and Donelson 1995; Djikeng et al. 1998). About 4000 of these sequences are derived from a cDNA library of a bloodstream *T. brucei* clone, and the remaining 500 are derived from a cDNA library of procyclic (insect form) organisms (E. Ullu, personal communication). Much information has been obtained from these EST sequences, which are available in the dbEST division of Genbank (http://www.ncbi.nlm.nih.gov/dbEST/index.html). About 15% of the ESTs are similar to a known gene of at least one other organism deposited in GenBank and result in the discovery of that gene in *T. brucei*. About 50% of the ESTs have no significant similarities to sequences in the databases, many of which likely represent trypanosomatid-unique genes. More than 500 ESTs have now been assigned to PFG-separated chromosomes of *T. brucei* stock TREU 927/4 (Melville et al. 1998) and have been mapped to clones of an arrayed P1 genomic DNA library.

Genomic DNA sequencing

A large-scale systematic sequencing of the *T. brucei* genome is currently underway at two sequencing centers: The Institute for Genomic Research (TIGR) in the United States and the Sanger Centre in the United Kingdom. *Trypanosoma brucei brucei* clone TREU 927 GUTat 10.1 expressing a specific VSG and derived from stock TREU (Trypanosomiasis Research Edinburgh University) 927/4 (GPAL/KE/70/EATRO1534) was selected as the reference strain for this sequencing effort because it (i) can be successfully passaged through tsetse flies in the laboratory, (ii) undergoes genetic exchange with other trypanosome isolates when they are co-infected into tsetse flies, (iii) grows well in culture as procyclic organisms, and (iv) can be readily transfected with foreign DNA.

The two sequencing centers have divided between themselves the specific megabase chromosomes to sequence, and the sequence determinations of the 1.15-Mb chromosome I (by the Sanger Centre) and the 1.3-Mb chromosome II (by TIGR) are nearly complete. In addition, both sequencing centers have conducted, or are conducting, end-sequence determinations of several thousand clones of *T. brucei* GUTat 10.1 genomic DNA, with the result that more than 20 Mb of genomic DNA sequence is currently available for searching for specific sequences. These randomly located DNA sequences, plus the sequences currently available for chromosomes I and II and portions of other chromosomes, can be downloaded from the web sites of the two sequencing centers: http://www.tigr.org/tdb/mdb/tbdb/ and http://www.sanger.ac.uk/Projects/T_brucei/.

The complete sequence of the *T. brucei* genome should be determined in the next few years, and will open the way for many research projects on African trypanosomes to move much more quickly and cheaply from ideas to experimental analyses of the ideas. A complete set of *VSG*s will be available for further elucidation of the still poorly understood mechanisms regulating antigenic variation that are at the heart of these parasites' unique ability to survive their hosts' immune responses. Genes in the nucleus can be screened for those encoding proteins that catalyze the unique phenomenon of kinetoplast RNA editing. Microarray technology, large–scale gene knockouts, proteomic approaches and other high-throughput methods can be applied to identify genes crucial to other aspects of the survival of the parasite in both the insect vector and the mammalian host. The end result, one can hope, is an efficient, cost-effective and better way to control and even eliminate this scourge of much of Africa.

REFERENCES

Alarcon C.M., M. Pedram, J.E. Donelson. 1999. Leaky transcription of variant surface glycoprotein gene expression sites in bloodstream African trypanosomes. J. Biol. Chem. **274**:16884-93.

Baltz T., C. Giroud, D. Baltz, C. Roth, A. Raibaud, and H. Eisen. 1986. Stable expression of two variable surface glycoproteins by cloned *Trypanosoma equiperdum*. Nature **319**:602-4.

Barry J.D., S.V. Graham, M. Fotheringham, V.S. Graham, K. Kobryn, B. Wymer. 1998. VSG gene control and infectivity strategy of metacyclic stage *Trypanosoma brucei*. Mol. Biochem. Parasitol. **91**:93-105.

Borst P., W. Bitter, P.A. Blundell, et al. 1998. Control of VSG gene expression sites in *Trypanosoma brucei*. Mol Biochem Parasitol **91**:67-76.

Borst P., L.H. van der Ploeg, J.F. van Hoek, J. Tas, and J. James. 1982. On the DNA content and ploidy of trypanosomes. Mol. Biochem. Parasitol. **6**:13-23.

Bresslau E., and L. Scremin. 1924. Die kerne der Trypanosomen und ihre verhalten zur nuclealreaktion. Archiv. protistenk. **10**:509-15.

C. elegans Sequencing Consortium. 1998. Genome sequence of the nematode *C. elegans*: a platform for investigating biology. Science **282**:2012-8.

Chen, K.K., and J.E. Donelson. 1980. The sequences of two kinetoplast minicircle DNAs of *Trypanosoma brucei*. Proc. Natl. Acad. Sci. USA **77**:2445-9.

Cross G.A.M, L.E. Wirtz, and M. Navarro. 1998. Regulation of *vsg* expression site transcription and switching in *Trypanosoma brucei*. Mol. Biochem. Parasitol. **91**:77-91.

Davies K.P., V.B. Carruthers, and G.A.M. Cross. 1997. Manipulation of the *vsg* co-transposed region increases expression-site switching in *Trypanosoma brucei*. Mol. Biochem. Parasitol. **86**:163-77.

Djikeng A., C. Agufa, J.E. Donelson, and P.A. Majiwa. 1998. Generation of expressed sequence tags as physical landmarks in the genome of *Trypanosoma brucei*. Gene **221**:93-106.

Donelson, J.E. 1996. Genome research and evolution in trypanosomes. Curr. Opin. Genet. Dev. **6**:699-703.

Donelson J.E., K.L. Hill, and N.M. El-Sayed. 1998. Multiple mechanisms of immune evasion by African trypanosomes. Mol. Biochem. Parasitol. **91**:51-66.

Donelson J.E., P.A.O. Majiwa, and R.O. Williams. 1979. Kinetoplast DNA minicircles of *Trypanosoma brucei* share regions of sequence homology. Plasmid **2**:572-88.

El-Sayed N.M., C.M. Alarcon, J.C. Beck, V.C. Sheffield, and J.E. Donelson. 1995. cDNA expressed sequence tags of *Trypanosoma brucei rhodesiense* provide new insights into the biology of the parasite. Mol. Biochem. Parasitol. **73**:75-90.

El-Sayed N.M., and J.E. Donelson. 1997. A survey of the *Trypanosoma brucei rhodesiense* genome using shotgun sequencing. Mol. Biochem. Parasitol. **84**:167-78.

El-Sayed N.M., P. Hegde, J. Quackenbush, S.E. Melville, and J.E. Donelson. 2000. The African trypanosome genome. Intern. J. Parasitol. **30**:329-45.

Ersfeld K., S.E. Melville, and K. Gull. 1999. Nuclear and genome organization of *Trypanosoma brucei*. Parasitol. Today **15**:58-63.

Esser K.M., and M.J. Schoenbechler. 1985. Expression of two variant surface glycoproteins on individual African trypanosomes during antigen switching. Science **229**:190-3.

Gibson, W.C., M. Crow, and J. Kearns. 1997. Kinetoplast DNA minicircles are inherited from both parents in genetic crosses of *Trypanosoma brucei*. Parasitol. Res. **83**:483-8.

Gibson W.C., and L. Garside. 1991. Genetic exchange in *Trypanosoma brucei brucei*: variable chromosomal location of housekeeping genes in different trypanosome stocks. Mol. Biochem. Parasitol. **45**:77-89.

Gibson W.C., L. Garside, and M. Bailey. 1992. Trisomy and chromosome size changes in hybrid trypanosomes from a genetic cross between *Trypanosoma brucei rhodesiense* and *T. b. brucei*. Mol. Biochem. Parasitol. **51**:189-99.

Guilbride, D.L., and P.T. Englund. 1998. The replication mechanism of kinetoplast DNA networks in several trypanosomatid species. J. Cell Science **111**:675-80.

Hasan G., M.J. Turner, and J.S. Cordingley 1984. Complete nucleotide sequence of an unusual mobile element from *Trypanosoma brucei*. Cell **37**:333-41.

Hope M., A. MacLeod, V. Leech, S. Melville, J. Sasse, A. Tait, and C.M. Turner. 1999. Analysis of ploidy (in megabase chromosomes) in *Trypanosoma brucei* after genetic exchange. Mol. Biochem. Parasitol. **104**:1-9.

Horn D., and G.A.M. Cross. 1997. Position-dependent and promoter-specific regulation of gene expression in *Trypanosoma brucei*. EMBO J. **16**:7422-31.

Kanmogne G.D., M. Bailey, and W.C. Gibson.1997. Wide variation in DNA content among isolates of *Trypanosoma brucei* ssp. Acta Trop. **63**:75-87.

Kimmel B.E., O.K. ole-MoiYoi, and J.R. Young. 1987. Ingi, a 5.2-kb dispersed sequence element from *Trypanosoma brucei* that carries half of a smaller mobile element at either end and has homology with mammalian LINEs. Mol. Cell. Biol. **7**:1465-75.

Lenardo M.J., K.M. Esser, A.M. Moon, L.H. Van der Ploeg, and J.E. Donelson. 1986. Metacyclic variant surface glycoprotein genes of *Trypanosoma brucei* subsp. *rhodesiense* are activated *in situ*, and their expression is transcriptionally regulated. Mol. Cell. Biol. **6**:1991-7.

Lips S., P. Revelard, and E. Pays. 1993. Identification of a new expression site-associated gene in the complete 30.5 kb sequence from the AnTat 1.3A variant surface protein gene expression site of *Trypanosoma brucei*. Mol. Biochem. Parasitol. **62**:135-7.

Mair G., H. Shi, H. Li, A. Djikeng, H.O. Aviles, J.R. Bishop, F.H. Falcone, C. Gavrilescu, J.L. Montgomery, M.I. Santori, L.S. Stern, Z. Wang, E. Ullu, and C. Tschudi. 2000. A new twist in trypanosome RNA metabolism: cis-splicing of pre-mRNA. RNA **6**:163-9.

Melville S.E. 1997. Genome research in African trypanosomes: chromosome size polymorphism and its relevance to genome mapping and anlaysis. Trans. Royal Soc. Trop. Med. Hyg. **91**:116-120.

Melville S.E., C.S. Gerrard, and J.M. Blackwell. 1999. Multiple causes of size variation in the diploid megabase chromosomes of African trypanosomes. Chromosome Res. **7**:191-203.

Melville S.E., V. Leech, C.S. Gerrard, A. Tait, and J.M. Blackwell. 1998. The molecular karyotype of the megabase chromosomes of *Trypanosoma brucei* and the assignment of chromosome markers. Mol. Biochem. Parasitol. **94**:155-73.

Mewes H.W., K. Albermann, M. Bahr, et al. 1997. Overview of the yeast genome. Nature **387**:7-65.

McDonagh P.D., P.J. Myler, and K.D. Stuart. 2000. The unusual gene organization of *Leishmania major* chromosome 1 may reflect novel transcription processes. Nucleic Acids Res. **28**:2800-3.

Muñoz-Jordán J.L., K.Pl Davies KP, and G.A.M. Cross. 1996. Stable expression of mosaic coats of variant surface glycoproteins in *Trypanosoma brucei*. Science **272**:1791-4.

Murphy N.B., A. Pays, P. Tebabi, et al. 1987. *Trypanosoma brucei* repeated element with unusual structural and transcriptional properties. J. Mol. Biol. **195**:855-71.

Myler, P.J, L. Audleman, T. deVos, G. Hixson, P. Kiser et al. 1999. *Leishmania major* Friedlin chromosome 1 has an unusual distribution of protein-coding genes. Proc. Natl. Acad. Sci. USA **96**:2902-6.

Navarro M., and G.A.M. Cross. 1996. DNA rearrangements associated with multiple consecutive directed antigenic switches in *Trypanosoma brucei*. Mol. Cell. Biol. **16**:3615-25.

Pays E., and D.P. Nolan. 1998. Expression and function of surface proteins in *Trypanosoma brucei*. Mol. Biochem. Parasitol. **91**:3-36.

Pays E., P. Tebabi, A. Pays, et al. 1989. The genes and transcripts of an antigen gene expression site from *T. brucei*. Cell **57**:835-45.

Pedram M., and J.E. Donelson. 1999. The anatomy and transcription of a monocistronic expression site for a metacyclic variant surface glycoprotein gene in *Trypanosoma brucei*. J. Biol. Chem. **274**:16876-83.

Rudenko G., P.A. Blundell, A. Dirks-Mulder, R, Kieft, and P. Borst. 1995. A ribosomal DNA promoter replacing the promoter of a telomeric VSG gene expression site can be efficiently switched on and off in *T. brucei*. Cell **83**:547-53.

Rudenko G., I. Chaves, A. Dirks-Mulder, and P. Borst. 1998. Selection for activation of a new variant surface glycoprotein gene expression site in *Trypanosoma brucei* can result in deletion of the old one. Mol. Biochem. Parasitol. **95**:97-109.

Schneider A., J. Martin, and N. Agabian. 1994. A nuclear encoded tRNA of *Trypanosoma brucei* is imported into mitochondria. Mol. Cell. Biol. **14**:2317-22.

Shapiro, T.A., and P.T. Englund. 1995. The structure and replication of kinetoplast DNA. Annl. Rev. Microbiol. **49**:117-43.

Sloof, P., A. de Haan, W. Eier, M. van Iersel, E. Boel, H. van Steeg, and R. Benne. 1992. The nucleotide sequence of the variable region in *Trypanosoma brucei* completes the sequence analysis of the maxicircle component of mitochondrial kinetoplast DNA. Mol. Biochem. Parasitol. **56**:289-300.

Sloof P., H.H. Menke, M.P. Caspers, and P. Borst. 1983. Size fractionation of *Trypanosoma brucei* DNA: localization of the 177- bp repeat satellite DNA and a variant surface glycoprotein gene in a mini-chromosomal DNA fraction. Nucleic Acids Res. **11**:3889-901.

Stuart K., T.E. Allen, M.L. Kable, and S. Lawson. 1997. Kinetoplastid RNA editing: complexes and catalysts. Curr. Opin. Chem. Biol. **1**:340-6.

Tait A., C.M. Turner, R.W. Le Page, and J.M. Wells. 1989. Genetic evidence that metacyclic forms of *Trypanosoma brucei* are diploid. Mol. Biochem. Parasitol. **37**:247-55.

Turner C.M. 1997. The rate of antigenic variation in fly-transmitted and syringe-passaged infections of *Trypanosoma brucei*. FEMS Microbiol. Lett. **153**:227-31.

Turner C.M., J.D. Barry, I. Maudlin, and K. Vickerman. 1988. An estimate of the size of the metacyclic variable antigen repertoire of *Trypanosoma brucei rhodesiense*. Parasitology **97**:269-76.

Turner, C.M., G. Hide, N. Buchanan, and A. Tait. 1995. *Trypanosoma brucei*: inheritance of kinetoplast DNA maxicircles in a genetic cross and their segregation during vegetataive growth. Exptl. Parasitol. **80**:234-41.

Turner C.M., S.E. Melville, and A. Tait. 1997. A proposal for karyotype nomenclature in *T. brucei*. Parasitol. Today **13**:5-6.

Van der Ploeg L.H., D. Valerio, T. De Lange, A. Bernards, P. Borst, and F.F. Grosveld. 1982. An analysis of cosmid clones of nuclear DNA from *Trypanosoma brucei* shows that the genes for variant surface glycoproteins are clustered in the genome. Nucleic Acids Res. **10**:5905-23.

Van der Ploeg L.H., D.C. Schwartz, C.R. Cantor, and P. Borst. 1984. Antigenic variation in *Trypanosoma brucei* analyzed by electrophoretic separation of chromosome-sized DNA molecules. Cell **37**:77-84.

Van der Ploeg L.H., C.L. Smith, R.I. Polvere, and K.M. Gottesdiener.1989. Improved separation of chromosome-sized DNA from *Trypanosoma brucei*, stock 427-60. Nucleic Acids Res. **17**:3217-27.

Vanhamme L., and E. Pays.1998. Controls of the expression of the VSG in *Trypanosoma brucei*. Mol. Biochem. Parasitol. **91**:107-16.

van Leeuwen F., R. Kieft, M. Cross, and P. Borst. 1998. Biosynthesis and function of the modified DNA base beta-D-glucosyl- hydroxymethyluracil in *Trypanosoma brucei*. Mol. Cell. Biol. **18**:5643-51.

van Leeuwen F., R. Kieft, M. Cross, and P. Borst. 2000. Tandemly repeated DNA is a target for the partial replacement of thymine by beta-D-glucosyl-hydroxymethyluracil in *Trypanosoma brucei*. Mol. Biochem. Parasitol. **109**:133-145.

Vickerman K. 1997. Landmarks in trypanosome research. *In* Trypanosomiasis and Leishmaniasis: Biology and Control, G. Hide, J.C. Mottram, G.H. Coombs, and P.H. Holmes (eds.). CAB International, Wallingford, UK, p. 1-37.

Weiden M., Y.N. Osheim, A.L. Beyer, and L.H. Van der Ploeg. 1991. Chromosome structure: DNA nucleotide sequence elements of a subset of the minichromosomes of the protozoan *Trypanosoma brucei*. Mol. Cell. Biol. **11**:3823-34.

Wells J.M., T.D. Prospero, L. Jenni, and R.W. Le Page. 1987. DNA contents and molecular karyotypes of hybrid *Trypanosoma brucei*. Mol. Biochem. Parasitol. **24**:103-16.

Wilson K., R.L. Berens, C.D. Sifri, and B. Ullman. 1994. Amplification of the inosinate dehydrogenase gene in *Trypanosoma brucei gambiense* due to an increase in chromosome copy number. J. Biol. Chem. **269**:28979-87.

Ziemann H. (1898). Eine Methode der Doppelfärbung bei Flagellaten, Pilzen, Spirillen un Bakterien, sowie bei einigen Amöben. Zentralbl. Bakteriol. Parasitenkd. Infektionskr. **24**:945-55.

Zomerdijk J.C., R. Kieft, and P. Borst. 1992. A ribosomal RNA gene promoter at the telomere of a mini-chromosome in *Trypanosoma brucei*. Nucleic Acids Res. **20**:2725-34.

Zomerdijk J.C., M. Ouellette, A.L. ten Asbroek, et al. 1990. The promoter for a variant surface glycoprotein gene expression site in *Trypanosoma brucei*. EMBO J. **9**:2791-801.

TOWARDS A TRYPANOSOMIASIS VACCINE

Samuel J. Black[1], Noel B. Murphy[2] and Derek P. Nolan[3]
1 Department of Veterinary and Animal Sciences, University of Massachusetts, Paige laboratory, Amherst, MA 01003, USA.
2 Laboratory of Molecular Parasitology, ILRI, PO Box 30709, Nairobi, Kenya.
3 Laboratory of Molecular Parasitology, Institute of Molecular Biology and Medicine, 12 Rue des Profs. Jeener et Brachet, Gosselies, B-6041.

ABSTRACT

This essay examines efforts at vaccine development for trypanosomiasis. We contrast antibodies with other trypanocidal materials of host origin, concluding that the goal of the trypanosome vaccine hunter should be to identify conserved trypanosome antigens that elicit a trypanocidal or trypanostatic antibody response that can be boosted by natural infection. In the search for conserved and surface accessible trypanosome antigens, on-going research indicates that tomato lectin-purified trypanosome components, which include growth factor receptors, elicit host protective antibody responses.

Key words Antigenic variation, conserved growth-factor receptors, LDL, HDL, flagellar pocket glycoproteins, vaccine development.

INTRODUCTION

The goal of the vaccine developer is to place traps that will ensnare, kill and deconstruct a target pathogen in the human body, or in the bodies of other chosen animals. These traps are typically set by injection, or ingestion, of selected antigens of pathogen origin, or agents (recombinant proteins, anti-idiotypic antibodies) that mimic these materials, or, if protective antigens are proteins, DNA encoding them. The traps are constructed from renewable resources of the host, namely a combination of specific lymphocytes, their products and a variety of accessory cells and molecules. The best ones are robust, operating for years after vaccination without further attention. The antigens used to set the traps need to be present in all strains of the target pathogen, accessible on their surface, and on the surface of cells harboring the pathogen if such are relevant, and capable of inducing immune responses that cripple the pathogen or its host cell, or both. In the absence of unequivocal acquired immunity to a disease, as is the case with African trypanosomiasis, the vaccine hunter must take an explorative approach. This entails the elucidation of the habitat and habits of the target pathogen, its vulnerabilities, conserved antigens that elicit immune responses that expose or exploit these vulnerabilities, the characteristics of appropriate and inappropriate immune responses, and the best ways to induce and sustain protective responses. The development of a vaccine against trypanosomiasis is a work in progress, the nature of which is discussed below.

THE TRYPANOSOME HABITAT

The human sleeping sickness parasites, *Trypanosoma brucei gambiense* and *T. b. rhodesiense*, live and replicate in the aqueous compartment of blood, lymph, interstitial fluids and, eventually, cerebrospinal fluids in mammals, as does *T. b .brucei*, one of the causative agents of Nagana. Intracellular forms of *T. brucei* have been detected in cultures containing nervous system tissue (Stoppini *et al.*, 2000) and in ependymal cells of mouse and rat brain (Abolarin *et al.*, 1982; Ormerod and Hussein, 1986). However, these are few in comparison to the extracellular forms. Intracellular forms of *T. brucei* may contribute to neuropathology associated with human sleeping sickness. They may also contribute to the maintenance of infection in patients and animals treated with chemotherapeutic agents, a role that is usually ascribed to trypanosomes present in cerebrospinal fluid.

Trypanosoma congolense and *T. vivax*, which together with *T. b. brucei* cause Nagana, also occupy the blood plasma and occasionally lymph and interstitial fluids. In addition to circulating in the blood, *T. congolense* adheres to blood vessel walls. Indeed the majority of *T. congolense* in an infected animal are in association with the walls of small blood vessels (Banks, 1978). Analyses *in vitro* suggest that attachment is to vascular endothelium via lectin like domains localized at distinct sites on the trypanosome flagellum (Hemphill *et al.*, 1994; Shakibaei *et al.*, 1994; Hemphill and Ross, 1995). To our knowledge there have been no reports of intracellular forms of *T. congolense* and *T. vivax*.

The presence of *T. brucei*, *T. congolense* and *T. vivax* in lymph and interstitial fluids in addition to blood plasma indicates a capacity to cross blood vessel, or micro blood vessel walls. The steps involved in this process are not documented in any detail, but presumably involve binding to endothelial cells and penetration between cellular junctions, or through cells. In regard to the latter issue, analyses of *T. brucei* extravasation in the brain suggest this occurs without disruption of the tight junction-specific proteins, occludin and zonula occludens 1 (Mulenga *et al.*, 2001). However, this does not preclude their passage between adjacent cells. Neutrophils and monocytes, whose directed diapedesis has been thoroughly characterized at the microscopic level, also pass between cells without permanent disruption of tight junctions (Edens and Parkos, 2000; Huber *et al.*, 2000). There may be considerable value in characterizing the binding aspect of the extravasation process used by trypanosomes. Inhibition of this process, if it could be achieved by immunological means, *e.g.*, by antibodies, would restrict the parasites to the vascular system. This would deny them the more voluminous space of the interstitial fluids while, at the same time, increase their exposure to antibodies, which are present at higher concentration in plasma than in the interstitial spaces.

NUTRIENTS AND GROWTH FACTORS

The African trypanosomes rely on their mammalian hosts for essential nutrients and growth factors. These are available in their fluid environment and include glucose, purines, low and high density lipoproteins (Black and Vandeweerd, 1989), and iron-loaded transferrin (Black and Vandeweerd, 1989; Schell *et al.*, 1991). Of the macromolecular growth factors/nutrients, LDL and HDL are needed throughout the first gap phase (G_1) of the trypanosome cell division cycle (Morgan *et al.*, 1993, 1996). Furthermore, exponentially growing bloodstream stage trypanosomes *in vitro* lack any store that is capable of sustaining G_1 progression in the absence of lipoproteins in medium. Hence, the parasites have a continuous need for this material (Morgan *et al.*, 1996). In addition, they irreversibly commit to a death process at $37°C$ when deprived of serum lipoproteins for a period of minutes to a few hours, depending on the stage of G_1 at which they are stopped (Morgan *et al.*, 1993, 1996). In contrast, the organisms can make several divisions in the absence of transferrin, presumably relying on stored Fe^{++} (Schell *et al.*, 1991; Morgan *et al.*, 1993).

Lipoprotein-induced *T. brucei* G_1 progression is independent of Ca^{++}, requires trypanosome protein synthesis and is inhibited by tyrphostin 51, which is an inhibitor of the epidermal growth factor (EGF) receptor tyrosine protein kinase. However, EGF does not substitute for lipoproteins to drive *T. brucei* G_1 progression. Furthermore, G_1 progression-associated tyrosine phosphorylation of *T. brucei* polypeptides is not apparent, and the precise lipoprotein ligand and putative trypanosome receptor interaction that stimulates

G_1 progression is not known (Morgan *et al.*, 1996). There is many a mile to walk on this trail before reaching its end. Trypanosome receptors that are used to acquire macromolecular nutrients and the endocytic process are discussed in detail by Nolan *et al.*, (2001) and will be considered later in this essay in the context of protective immune responses.

Lipoproteins are available in lymph and interstitial fluids as well as in blood plasma (Raymond *et al.*, 1985; Reichl, 1990; Oliveira *et al.*, 1993; Jaspard *et al.*, 1997; Danik *et al.*, 1999; Nanjee *et al.*, 2001). However, the variety of density classes present, as well as their apoprotein and lipid contents, may differ among these sites. For example, very low density lipoproteins are absent from lymph (Reichl 1990). This absence may be of little significance to trypanosomes. In this regard, two strains of *T. brucei,* were unable to utilize this class of lipoprotein for growth *in vitro* (Black and Vandeweerd, 1989). The lipoproteins of cerebrospinal fluid are analogous to those in plasma but have a different cellular origin (Ladu *et al.*, 2000). The impact of lipoprotein diversity on the replication of trypanosomes in different niches within a host is unknown. In the case of *T. brucei*, replicating forms are often more prevalent in lymph than in blood (Ssenyonga and Adam, 1995).

TRYPANOCIDAL AND TRYPANOSTATIC AGENTS

When required host components are available in adequate quantity, African trypanosomes undergo repeated cycles of mitosis (Vandeweerd and Black, 1989; Morgan *et al.*, 1996), until this is stopped by endogenous and host-derived growth-inhibitors (Seed and Black, 1999; reviewed in Murphy and Olijhoek, 2001), or the parasites are killed by antibodies (Macaskill *et al.*, 1980; Crowe *et al.*, 1984; Mclintock *et al.*, 1993), or both. Peak levels of *T. brucei* parasitemia correspond to the accumulation of growth-incompetent parasites in the blood. These are in the first gap phase (G_1) of the cell division cycle and are infective for tsetse flies. The kinetics of *T. brucei* G_1 arrest can be similar in intact and irradiated mice that cannot make VSG-specific antibodies, or respond to third party antigens (Black *et al.*, 1985). Furthermore, even in the most resistant strains of mice, peak levels of parasitemia are less than 5 fold higher in immunocompromized as compared to intact mice, and in the latter, antibody-dependent clearance occurs after the majority of blood parasites are arrested in G_1 (Black *et al.*, 1985) suggesting that antibody plays a more significant role in clearance of *T. brucei* in mice, than in determining the overall level of parasitemia. It is not presently known whether trypanosome G_1 arrest in immunocompromized mice results from feedback inhibition via a trypanosome product, as suggested by the work of Murphy and Olijhoek (2001), or host-derived material that is induced in a trypanosome-density dependent manner.

The extent to which trypanosome G_1 arrest contributes to control of parasitemia in mammal species other than mice is unclear. Trypanosome-infected bovids have much lower levels of parasitemia than arise in infected mice, as discussed by Naessens *et al.*, (2001), and some African wildlife species, that are naturally selected to be resistant to trypanosomiasis, suppress parasitemia after only one or a few parasitemic waves, indicating the involvement of an acquired response (Reduth *et al.*, 1994; Wang *et al.*, 1999). Furthermore, the role of antibody-independent control of *T. congolense* and *T. vivax* parasitemia in mice is much less clear than with *T. brucei*. Indeed, experiments with a cloned *T. vivax* showed that clearance of the first parasitemic wave occurred while the parasites were in exponential growth and was mediated by antibody (Mahan *et al.*, 1986, 1989).

Clearance of trypanosomes at peak parasitemia is mediated by antibodies specific for exposed epitopes on trypanosome-attached variable surface glycoprotein (VSG; Black *et al.*, 1985). The expressed VSG on any trypanosome is encoded by a single VSG gene, that is switched among a repertoire of hundreds of VSG genes and pseudogenes. In addition, trypanosomes have an error prone polymerase and by recombination generate new VSG genes. Variation of the VSG coat among clonal progeny of trypanosomes, allows some organisms to evade immune elimination. Growth of the VSG variants after dominant VSG types are cleared by antibody in each parasitemic wave results in a further wave of trypanosomes, a process that continues until the death of infected susceptible hosts

Once an effective concentration is achieved, VSG-specific antibodies are highly efficient at removing target trypanosomes. However, a broader acting agent is required to terminate infection. Several serum components, in addition to VSG-specific antibody, have been shown to kill trypanosomes, and to do so in a VSG-independent manner. These include: human serum very high density lipoprotein and a non-lipoprotein complex of haptoglobin related protein and IgM, which lyse *T. b. brucei*, *T congolense* and *T.vivax* but not *T.b.gambiense,* or *T.b. rhodesiense* (reviewed in Black *et al.*, 2001a). Heat-sensitive serum components, putatively complement components, that have a restricted mammal species distribution, can also kill *T. b. brucei* (Black *et al.*, 1999) but have not been tested on different isolates of this trypanosome subspecies, or on other trypanosome species. Recombinant mouse tumor necrosis factor alpha (TNF α) has been reported to kill *T. b. brucei in vitro* (Magez *et al.*, 1997) although this effect appears to be restricted to only some preparations and very high concentrations are required (Kitani, Nakamura, Naessens, Murphy and Black, unpublished data). Knock-out of the gene encoding TNF α results in elevated parasitemia in mice (Magez *et al.*, 1999) consistent with either a direct effect of TNF α on bloodstream stage trypanosomes, or a downstream effect involving TNF α-induced material. With regard to the latter possibility, TNF α stimulates phospholipases that cleave fatty acids from membrane phospholipids. Should this generate significant quantities of medium-chain length fatty acids, trypanosome cell cycle progression will be inhibited (Morgan *et al.*, 1993). Reactive nitrogen and oxygen intermediates also have trypanocidal activity *in vitro*. However, the former are unlikely to exert a significant effect in the bloodstream because of their inactivation by reaction with hemoglobin (Mabbott *et al.*, 1994). Furthermore, knockout of the host gene encoding the inducible NO synthase does not affect trypanosome parasitemia, at least in mice (Millar *et al.*, 1995). Of the reactive oxygen intermediates, hydrogen peroxide has the most potent impact on the parasites, but is typically quenched by serum catalase limiting its involvement in control of the blood-dwelling parasites (Wang *et al.*, 1999). An exception to this has been noted in Cape buffalo, which are naturally selected to be resistant to trypanosomiasis. Trypanosome-infected Cape buffalo transitorily develop a 5 to 8 fold reduction in their already low serum catalase activity, which allows the generation of trypanocidal H_2O_2 during the catabolism of endogenous purine by serum xanthine oxidase (Wang *et al.*, 1999; Wang *et al.*, 2001) resulting in non-specific killing of trypanosomes. This may contribute to the rapid control of acute phase parasitemia in trypanosome-infected Cape buffalo (Wang *et al.*, 1999).

TNF α, reactive oxygen intermediates, and reactive nitrogen intermediates are short-range effectors in the immune system and the high concentrations required to kill trypanosomes in blood plasma would more likely contribute to pathology rather than cure. It is a fair bet that these trypanocidal/static agents are peripheral to the true game of cat and mouse that is played by trypanosomes and immune systems. Although *T. brucei* contain material that induces release of INF γ from CD8 positive T cells (Bakhiet *et al.*, 1993; Vaidya *et al.*, 1997), there is no evidence that they induce these cells to cytotoxic function, or indeed that non-specifically activated CD8 T-cells would kill trypanosomes. African trypanosomes have been shown to be relatively insensitive to the actions of NK cells and isolated cytolysincontaining granules (Albright *et al.,* 1988) suggesting that cell contact-dependent killing processes are likely to play little or no role in defense against these parasites. To our knowledge, data are not available on the impact of host-derived microbicidal peptides, *e.g.,* defensins, on African trypanosomes, but it seems unlikely that these could reach an adequate concentration in plasma to damage the eukaryotic parasites without also affecting host cells.

Based on the above considerations, it is our opinion that non-specific defenses that become elevated during infection are unlikely to be major players in the control of trypanosome parasitemia. In support of this contention, it has been shown that many of these non-specific defenses are highly elevated in trypanosome-infected mice, especially at the time of wave remission (Grosskinsky *et* al., 1983; Daulouede *et* al., 1994; Dumas and Bouteille, 1996; Gobert *et* al., 1998; Sternberg *et* al., 1998; Wang *et al.,* 1999) but are inadequate to cause elimination of the infection. Indeed, innate immune responses often appear to exert pathologic and immunosuppressive effects, rather than preventing subsequent wave development, which occurs within a few days of clearance of the previous wave. While differences in cytokine responses, and oxidative or other non-specific responses in infected animals affect parasite-induced host pathology and host survival, issues that are discussed by Mansfield *et al.,* (2001) and Naessens *et al.,* (2001), these differences in host response to infection are not of primary concern to the vaccine hunter, whose intent is to prevent infection, or cause its rapid termination thus preventing infection-associated pathology.

ONLY AN ANTIBODY BASED VACCINE WILL DO

Antibodies in combination with complement factors and macrophages kill trypanosomes and do so at physiologically relevant concentrations. Antibodies of the IgG class are relatively long lived and are generated by T-cell-dependent B-lymphocytes, which display immunological memory and the capacity for a rapid recall response upon restimulation. Clearly, a vaccine that could induce a host protective antibody response would be highly desirable. Several tacks are being taken towards this goal. An effective trypanosome vaccine might, for example, induce a VSG-independent trypanocidal antibody response that can be restimulated after infection. This is the goal we are working towards. Alternatively, a vaccine that caused immune responses against VSG antigens to be accelerated and amplified to the extent that parasites are killed while present at a low number, would also be desirable. One way of achieving this might be to neutralize trypanosome-induced immunosuppression, which is considered to be responsible for poor parasite-specific responses, by generating prophylactic responses against the putative "trypanosuppressin". This goal was championed in the 1980s by Askonas and her group, and has been frequently re-visited by scientists who are investigating the molecular basis of immunodepression (reviewed in Mansfield *et al.*, 2001; Naessens *et al.*, 2001). Another way might be to induce TH2 responses against conserved amino acids sequences on VSGs as discussed below.

CONSERVED TRYPANOSOME ANTIGENS
VSG-based vaccines

Infection of trypanosomiasis-susceptible mammals with cloned trypanosomes, or their immunization with lethally-irradiated trypanosomes, elicits high titer antibody-responses against the immunodominant VSGs of the parasites. VSG-specific antibodies are effective at killing the parasites to which they were raised but are ineffective at killing VSG-different clones of trypanosomes. Although there are groups of conserved amino acids in VSGs that are responsible for turns in sub-regions of the protein (Reinitz *et al.,* 1992), these are not accessible to antibody on intact trypanosomes. And, while the VSG repertoire expressed by metacyclic trypanosomes, i.e., the infective form transmitted from tsetse flies to mammals, is much smaller than that of the bloodstream stage parasites (Turner *et* al., 1988; Barry *et* al., 1998), it is unstable in the case of *T. b. rhodesiense* (Barry *et al.,* 1983), and most likely in the case of the other Salivarian trypanosomes.

Because of their high degree of polymorphism, buried conserved elements, and minimal conserved sequence among unrelated VSGs, these molecules do not induce broad-acting host protection. For this reason, efforts to develop VSG-based vaccines have been abandoned. However, it would be of interest to investigate whether TH2 T-cell immune responses could be induced against recombinant polypeptides that are engineered to contain the conserved amino acid sequences of related VSGs and none of the variable sequences. Here we are envisaging that the primed TH2 T-cells might deliver "go faster" and "IgG class switch" signals to B-cells specific for exposed epitopes on trypanosome-attached VSG in infected animals. We anticipate that Ig receptors on such B-cells bind to GPI-anchored VSGs on their target trypanosome, and that the VSGs are plucked from the surface of the writhing organisms, endocytosed, processed, and their peptides presented back on the B-cell surface in association with class II MHC antigens ready to engage the peptide/MHCII specific TCRs of the TH2 T cells. A fantastic scenario perhaps, but not an impossible one, unless the energy of VSG-homodimer associations, and of GPI and membrane-phospholipid associations, preclude the plucking of VSG as shearing forces detach specific B-cells from the surface of the parasites.

Surface exposed non-VSG antigens

Characterization of the surface of living trypanosomes by lectin-binding (Balber and Frommel, 1988; Brickman and Balber, 1990), enzyme-catalyzed labeling procedures (Jackson et al., 1993; Nolan et al., 1997) and biotinylation (Ziegelbauer and Overarth, 1992) as well as heterologous expression of trypanosomal cDNAs in COS cells (Nolan et al., 2000) has revealed invariant surface components in addition to VSG. These externally-disposed conserved components are distributed over the parasite surface, or in surface microdomains including the flagellar pocket (FP). Most of the conserved external-disposed molecules that have been detected on bloodstream stage trypanosomes are developmentally regulated, consistent with their fulfilling a stage-specific function. However, none of these have so far been found to elicit host-protective antibodies. In addition these proteins do not appear to be accessible to antibodies in live cells and could be localized only using fixed trypanosomes where packing of the VSG coat was disrupted. This finding suggests that the VSG coat acts as a shield and covers invariant proteins that potentially might be targeted by the antibody response. The one place this shield must be compromised is the FP where the binding of required macromolecules to receptors, and shielding by VSG, are unlikely to be simultaneously compatible.

Several investigators have focused on the vaccine potential of antigens of the trypanosome flagellar pocket (FP) (McLaughlin, 1987; Olenick et al.,1988; Mkunza et al., 1995; Radwanska et al., 2000), which is a flask shaped domain of the plasma membrane that is both specialized for, and the sole site of, endocytosis and exocytosis (reviewed in Nolan et al., 2001). The flagellum emerges from and partially occupies the FP, and its beating may facilitate exchange of the pocket contents with the extracellular fluids. The FP contains receptors for host-derived nutrients and growth factors (Pays and Nolan, 1998) and may sequester these from the immune system so that they do not typically induce antibody responses. Protection against low dose challenge with some *T. brucei* has been achieved in mice by immunization with FP membrane antigens (McLaughlin, 1987; Olenick et al.,1988; Mkunza et al., 1995; Radwanska et al., 2000). However, immunization of mice with defined trypanosome FP receptors has met with little success. In this regard, antibodies against the trypanosome receptor for transferrin were found to affect trypanosome replication, but only when the affinity of interaction between transferrin and this polymorphic receptor was low (Borst *et al.*, 1996). Antibodies against a putative 145 kDa *T. brucei* LDL receptor that is associated with the FP have also been shown to inhibit the growth of a bloodstream stage *T. brucei* clone *in vitro* (Coppens *et al.*, 1988). However, the antibodies did not affect other cloned bloodstream stage *T. brucei in vitro* (discussed in Black *et al.*, 2000) and immunization of mice with the receptor material had little, or no, effect on their lack of capacity to control trypanosomes (Bastin *et al.*, 1996). The identity of the 145 kDa material as an LDL receptor has been called into doubt in recent years, and on-going studies indicate that the 145 kDa material is in fact a fragment of the Apo B component of LDL and not of trypanosome origin (Opperdoes, personal communication).

Tomato lectin-binding trypanosome antigens and growth-inhibitory antibodies

Reliable induction of immunity might require targeting of specific epitopes on trypanosome receptors, or the generation of antibodies against a combination of receptors. However, difficulties in obtaining adequate amounts and diversity of such molecules for immunization studies, have limited exploitation of this approach until now. The recent discovery (Nolan *et al.*, 1999) that trypanosome FP-restricted glycoproteins bear N-linked glycans containing linear poly-N-acetyllactosamine (pNAL) provides a straightforward means for their isolation and thus, their immunological evaluation. The linear pNAL moiety is selectively bound by immobilized tomato lectin (TL) and bound material can be eluted by chito-oligosaccharides, and most effectively by tri-N-acetyl-chitotriose and tetra-N-acetyl-chitotetraose both of which are highly represented in chitin hydrolysate. TL-chromatography reproducibly results in the isolation of approximately 35 *T. brucei* proteins and associated molecules as determined by 2D gel analysis, silver staining and metabolic labeling with ^{35}S cysteine and methionine (Nolan, unpublished). This fraction represents about 0.2% of the toal cellular protein solubilized by CHAPS. This TL-binding trypanosome material includes components that bind growth-inhibitory IgG that arises naturally in infected Cape buffalo and that may control trypanosome population growth during the chronic phase of infection (discussed in Black *et al.*, 2001). Immunization of mice with the TL-isolated material induced trypanosome growth-inhibitory serum activity, that was detected *in vitro* and was neither trypanosome strain- nor species-restricted (unpublished data). These mice also acquired resistance to infection with a limited dose of *T. congolense* and to a lesser extent *T. brucei*. Immunization with 20 µg glycopeptidase F-treated TL-binding material from *T. brucei* ILTat 1.1 and boosting twice after 2 and 4 weeks protected 6 of 7 mice from infection initiated by an intraperitoneal (ip) injection of 500 *T. congolense* IL1180, 3 of 7 mice from infection by an ip injection of 500 *T. brucei* GUTat 3.1 in one experiment and 2 of 8 mice from infection with the same parasites in a repeat experiment. It also protected 2 of 4 mice from infection by an ip injection of *T. brucei* S427 clone 22 (unpublished data). In these experiments, all the control mice, which had been primed and boosted with an irrelevant antigen in adjuvant, became infected.

Evaluation of immune sera in vitro

TL-binding trypanosome material includes ESAG 6 and 7 gene products, which jointly comprise the transferrin receptor (Salmon *et al.*, 1994; Steverding *et al.*, 1995), and the 145 kDa material that is now considered to a fragment of Apo B, suggesting the simultaneous presence of an LDL receptor. In addition to these components, the TL-binding material contains several other glycoproteins of unknown function, any, or all of which might be accessible to antibody. The growth-inhibitory antibodies may exert their effect by depriving trypanosomes of required growth factors/nutrients through competition for receptors, or may have a direct effect on the cells e.g., by reacting with molecules that transduce growth-inhibitory signals. Either process would account for the *in vitro* trypanosome growth-inhibitory activity of serum from mice that have been primed and boosted with TL-binding trypanosome material, and of IgG from infected Cape buffalo. The latter was shown to be specific for TL-binding trypanosome material by competitive inhibition (unpublished data).

In vitro trypanosome growth-inhibitory activity was not fully predictive of *in vivo* immunity. All mice that had been primed and boosted with TL-binding material from *T. brucei* ILTat 1.1 developed *T. brucei* S427 clone 22 growth-inhibitory activity in their sera, and this was detected to 10% vol. in growth medium. However, only 2 of the 4 mice tested resisted infection with *T. brucei* S427 clone 22. The development of parasitemia was delayed for a few days in 1 of the remaining 2 mice and unaffected in the other. These data are puzzling. Perhaps the protected mice continued to make antibodies that are capable of delivering a growth-inhibitory signal after infection, or had higher amounts of IgG that could effectively compete the binding of required serum growth factors to their trypanosome FP-restricted receptor(s) *in vivo*, or had lower concentrations of the relevant growth factors, or had a trypanocidal/static concentration of antibodies in lymph and interstitial fluids as well as in serum, or a combination of these. It should be possible to determine which, if any, of these possibilities is correct using monoclonal antibody techniques and more defined culture systems. It will also be possible to use more defined antigen preparations as the identities of the TL-binding trypanosome material are revealed using proteomic and genomic approaches.

Vaccination would be a beneficial addition to the tools presently available to control sleeping sickness and nagana. The protection studies that we have carried out suggest that we are on the right track to finding vaccine antigens. We believe that further research in this area is highly warranted and likely to be productive.

REFERENCES

Abolarin, M.O., D.A. Evans, D.G. Tovey, and W.E. Ormerod. 1982. Cryptic stage of sleeping-sickness trypanosome developing in choroids plexus epithelial cells. Br. Med. J. (Clin. Res. Ed.) **285**:1380-1382.

Albright, J.W., W.E. Munger, P.A. Henkart, and J.F. Albright. 1988. The toxicity of rat large granular lymphocyte tumor cells and their cytoplasmic granules for rodent and African trypanosomes. J. Immunol. **140**:2774-2778.

Bakhiet, M., E. Mix, K. Kristensson, H. Wigzell, and T. Olsson. 1993. T cell activation by a Trypanosoma brucei brucei-derived lymphocyte triggering factor is dependent on tyrosine protein kinases but not on protein kinase C and A. European Journal of Immunology **23**:1535.

Balber, A.E., and T.O.Frommel. 1988. *Trypanosoma brucei gambiense* and *T. b. rhodesiense* concanavalin A binding to the membrane and flagellar pocket of bloodstream and procyclic forms. J. Protozool. **35**: 214-219.

Banks, K.L. 1978. Binding of *Trypanosoma congolense* to the walls of small blood vessels. J. Protozool. **25**:241-245.

Barry, J.D., J.S. Crowe, and K. Vickerman. 1983. Instability of the *Trypanosoma brucei rhodesiense* metacyclic variable antigen repertoire. Nature **306**: 699-701.

-------------, S.V. Graham, M. Fotheringham, V.S. Graham, K. Kobryn, and B. Wymer. 1998. VSG gene control and infectivity strategy of metacyclic stage *Trypanosoma brucei*. Mol. Biochem. Parasitol. **91**: 93-105.

Bastin, P., A. Stephan, J. Raper, J.M. Saint-Remy, F.R. Opperdoes, and P.J. Courtroy. 1996. An M_r 145000 low-density lipoprotein (LDL)-binding protein conserved throughout the Kinetoplastida order. Mol. Biochem. Parasitol. **76**: 43-56.

Black, S.J., C.N. Sendashonga, C. O'Brien, N.K. Borowy, M. Naessens, P. Webster, and M. Murray. 1985. Regulation of parasitemia in mice infected with *Trypanosoma brucei*. Curr. Top. Microbiol. Immunol. **117**:93-115.

-----------, and V. Vandeweerd. 1989. Serum lipoproteins are required for multiplication of *T. brucei* under axenic culture conditions. Mol. Biochem. Parasitol. **37**: 65-72.

-----------, S.J., Q. Wang, T. Makadzange, Y.L. Li, A. Van Praagh, M. Loomis, and J.R. Seed. 1999. Anti-*Trypanosoma brucei* activity of non-primate zoo sera. J. Parasitol. **85**:48-53.

-----------, J.R. Seed, and N.B. Murphy. 2001. Innate and acquired resistance to African trypanosomiasis. J. Parasitol. **87**:1-9.

-----------, E.L. Siccard, N.B. Murphy, and D. Nolan. 2001a. Innate and acquired control of trypanosome parasitaemia in Cape buffalo. Int J Parasitol. **31**:561-564.

Borst, P., W. Bitter, P. Blundell, M. Cross, R. McCulloch, G. Rudenko, M.C. Taylor, and F. Van Leeuven. 1996. The expression sites for variant surface glycoproteins of *Trypanosoma brucei*. Biology and Control: British Soc Parasitol/CAB Int, Oxford: 109-131.

Brickman, M.J., and A.E. Balber. 1990. *Trypanosoma brucei rhodesiense* bloodstream forms: surface ricin-binding glycoproteins are localized exclusively in the flagellar pocket and the flagellar adhesion zone. J. Protozool. **37**: 219-224.

Coppens, I., P. Baudhuin, F.R. Opperdoes, and P.J. Courtroy. 1988. Receptors for the host low density lipoproteins on the hemoflagellate *Trypanosoma brucei*: purification and involvement in the growth of the parasite. Proc. Natl. Acad. Sci. U.S.A. **85**: 6753-6757.

Crowe, J.S., A.G. Lamont, J.D. Barry, and K. Vickerman. 1984. Cytotoxicity of monoclonal antibodies to *Trypanosoma brucei*. Trans. R. Soc. Trop. Med. Hyg. **78**:508-513.

Danik, M., D. Champagne, C. Petit-Turcotte, U. Beffert, and J. Poirier. 1999. Brain lipoprotein metabolism and its relation to neurodegenerative disease. Crit. Rev. Neurobiol. **13**:357-407.

Daulouede, P.S., M.C. Okomo-Assoumou, M. Labassa, C. Fouquet, and P. Vincendeau. 1994. Defense mechanisms in trypanosomiasis. Bull. Soc. Pathol. Exot. **87**:330-332.

Dumas, M., and B. Bouteille. 1996. Human African trypanosomiasis. C. R. Seances Soc. Biol. Fil. **190**:395-408.

Edens, H.A., and C.A. Parkos. 2000. Modulation of epithelial and endothelial paracellular permeability by leukocytes. Adv. Drug Deliv. Rev. **41**:315-328.

Gobert, A.P., S. Semballa, S. Daulouede, S. Lesthelle, M. Taxile, B. Veyret, and P. Vincendeau. 1998. Murine macrophages use oxygen- and nitric oxide-dependent mechanisms to synthesize S-nitroso-albumin and to kill extracellular trypanosomes. Infect. Immun. **66**:4068-4072.

Grosskinsky, C.M., R.A. Ezekowitz, G. Berton, S. Gordon, and B.A. Askonas. 1983. Macrophage activation in murine African trypanosomiasis. Infect. Immun. **39**:1080-1086.

Hemphill, A., I. Frame, and C.A. Ross. 1994. The interaction of *Trypanosoma congolense* with endothelial cells. Parasitology **109**:631-641.

---------------, and C.A. Ross. 1995. Flagellum-mediated adhesion of *trypanosoma congolense* to bovine aorta endothelial cells. Parasitol. Res. **81**:412-420.

Huber, D., M.S. Balda, and K. Matter. 2000. Occludin modulates transepithelial migration of neutrophils. J. Biol. Chem. **275**:5773-5778.

Jackson, D.G., H.J. Windle, and H.P. Voorheis. 1993. The identification, purification, and characterization of two invariant surface glycoproteins located beneath the surface coat barrier of bloodstream form *Trypanosoma brucei*. J. Biol. Chem. **268**: 8085-8095.

Jaspard, B., N. Fournier, G. Vietitez, V. Atger, R. Barbaras, C. Vieu, J. Manent, H. Chap, B. Perret, and X. Collet. 1997. Structural and functional comparison of HDL from homologous human plasma and follicular fluid. A model for extravascular fluid. Arterioscler. Thromb. Vasc. Biol. **17**:1605-1613.

Ladu, M.J., C. Reardon, L. Van Eldik, A.M. Fagan, G. Bu, D. Holtzman, and G.S. Getz. 2000. Lipoproteins in the central nervous system. Ann. N. Y. Acad. Sci. **903**:167-175.

Mabbott, N.A, I.A. Sutherland, and J.M. Sternberg. 1994. Trypanosoma brucei is protected from the cytostatic effects of nitric oxide under in vivo conditions. Parasitol. Res. **80**, 687-690

Macaskill, J.A., P.H. Holmes, D.D. Whitelaw, I. McConnell, F.W. Jennings, G.M. Urquhart. 1980. Immunological clearance of ^{75}Se-labelled *Trypanosoma brucei* in mice. II. Mechanisms in immune animals . Immunology **40**:629-635.

Magez S, M. Radwanska, A. Beschin, K. Sekikawa, and P. De Baetselier. 1999. Tumor necrosis factor alpha is a key mediator in the regulation of experimental Trypanosoma brucei infections. Infect Immun. **67**:3128-32.

----------, M. Geuskens, A. Beschin, H. del Favero, H. Verschueren, R. Lucas, E. Pays, and P. de Baetselier. 1997. Specific uptake of tumor necrosis factor-alpha is involved in growth control of Trypanosoma brucei. J Cell Biol. **137**:715-27.

Mahan, S.M., L. Hendershot, and S.J. Black. 1986. Control of trypanodestructive antibody responses and parasitemia in mice infected with *Trypanosoma (Duttonella) vivax.* Infect. Immun. **54**:213-221.

-----------, and S.J. Blacl, 1989. Differentiation, multiplication, and control of bloodstream form *Trypanosoma (Duttonella) vivax.* J. Protozool. **36**:424-428.

Mansfield, J.M., T.H. Davis, and M.E, Dubois. 2001. Immunobiology of African Trypanosomiasis: New paradigms, newer questions. *In* World Class Parasites, Vol. I. The African Trypanosomes. S.J. Black and J.R. Seed Eds. Pp. 79-96. Kluwer Academic Publishers.

McLaughlin, J. 1987. *Trypanosoma rhodesiense*: antigenicity and immunogenicity of flagellar pocket components. Exp. Parasitol. **64**: 1-11.

Mclintock, L.M., C.M. Turner, K. Vickerman. 1993. Comparison of the effects of immune killing mechanisms on *Trypanosoma brucei* parasites of slender and stumpy morphology. Parasite Immunol. **15**:475-480.

Millar, A.E., J. Sternberg, C. McSarry, X.Q. Wei, F.Y. Liew, and C.M. Turner. 1999. T-cell responses during infection in mice deficient in inducible nitric oxide synthase. Inf. Imm. **67**, 3334-3338.

Mkunza, F., W.M. Olaho, and C.N. Powell. 1995. Partial protection against natural trypanosomiasis after vaccination with a flagellar pocket antigen from *Trypanosoma brucei rhodesiense.* Vaccine **13**: 151-154.

Morgan, G.A., H.B. Laufman, F.P. Otieno-Omondi, and S.J. Black. 1993. Control of G1 to S cell cycle progression of *Trypanosoma brucei* S427 organisms under axenic culture conditions. Mol. Biochem. Parasitol. **57**: 241-252.

-----------------, E.A. Hamilton, and S.J. Black 1996. The requirements for G1 checkpoint progression of *Trypanosoma brucei* S 427 clone 1. Mol. Biochem. Parasitol. **78**: 195-207.

Mulenga, C., J.D. Mhlanga, K. Kristensson, and B. Robertson. 2001. *Trypanosoma brucei brucei* crosses the blood-brain barrier while tight junction proteins are preserved in a rat chronic disease model. Neuropathol. Appl. Neurobiol. **27**:77-85.

Murphy, N.B., and T. Olijhoek. 2001. Trypanosome factors controlling population size and differentiation status. *In* World Class Parasites, Vol. I. The African Trypanosomes. S.J. Black and J.R. Seed Eds. Pp. 113-126. Kluwer Academic Publishers.

Naessens, J., D.J. Grab, M. Sileghem. 2001. Identifying the mechanism of trypanotolerance in cattle. *In* World Class Parasites, Vol. I. The African Trypanosomes. S.J. Black and J.R. Seed Eds. Pp. 97-112. Kluwer Academic Publishers.

Nanjee, M.N., C.J. Cooke, J.S. Wong, R.L. Hamilton, W.L. Olszewski, and N.E. Miller. 2001. Composition and ultrastructure of size sub-classes of normal human peripheral lymph lipoproteins. Quantification of cholesterol uptake by HDL in tissue fluids. J. Lipid Res. **42**:639-648.

Nolan, D.P., J.A. Garcia-Salcedo, M. Geuskens, D. Salmon, F. Paturiaux-Hanocq, A. Pays, P. Tebabi, and E. Pays. 2001. Endocytosis in African trypanosomes. *In* World Class Parasites, Vol. I. The African Trypanosomes. S.J. Black and J.R. Seed Eds. Pp. 127-142. Kluwer Academic Publishers.

--------------, D.G. Jackson, M.J. Biggs, E.D. Brabazon, A. Pays, F. Van Laetham, F. Paturiaux-Hanocq, J.F. Elliot, H.P. Voorheis and E. Pays. 2000 Characterization of a novel alanine-rich protein located in surface microdomains in *Trypanosoma brucei*. J. Biol. Chem. **275**: 4072-4080.

--------------, M. Geuskens, and E. Pays. 1999. Linear poly-N-acetyllactosamine as sorting signals in exo/endocytosis in *Trypanosome brucei*. Current Biol. **9**:1169-1172.

--------------, D.G. Jackson, H.J. Windle, A. Pays, M. Geuskens, A. Michel, H.P. Voorheis, and E. Pays. 1997. Characterization of a novel, stage-specific, invariant surface protein in *Trypanosoma brucei* containing an internal serine-rich, repetitive motif. J. Biol. Chem. **272**: 29212-29221.

Olenick, J.G., R. Wolff, R.K. Nauman, and J. McLaughlin. 1988. A flagellar pocket membrane fraction from *Trypanosoma brucei rhodesiense*: immunogold localization and nonvariant immunoprotection. Infect. Immun. **56**: 92-98.

Oliveira, H.C., K. Nilausen, H. Meinertz, and E.C. Quintao. 1993. Cholesteryl esters in lymph chylomicrons: contribution from high density lipoprotein transferred from plasma into intestinal lymph. J. Lipid Res. **34**:1729-1736.

Ormerod, W.E., and M.S. Hussein. 1986. The ventricular ependyma of mice infected with *Trypanosoma brucei*. Trans. R. Soc. Trop. Med. Hyg. **80**:626-633.

Raymond, T.L., S.A. Reynolds, and J.A. Swanson. 1985. Lipoproteins of the extravascular space: enhanced macrophage degradation of low density lipoproteins from interstitial inflammatory fluid. J. Lipid Res. **26**:1356-1362.

Pays, E. and D.P. Nolan. 1998. Expression and function of surface proteins in *Trypanosoma brucei*. Mol. Biochem. Parasit. **91**: 3-36.

Radwanska, M., S. Magez, N, Dumont, A. Pays, D. Nolan, and E. Pays. 2000. Antibodies against the flagellar pocket fraction of *Trypanosoma brucei* preferentially recognize HSP60 in cDNA expression library. Parasite Immunol. **22**:639-650.

Reduth, D., J.G. Grootenhuis, R.O. Olubayo, M. Muranjan, F.P. Otieno-Omondi, G.A. Morgan, R. Brun, D.J.L. Williams, and S.J. Black. 1994. African buffalo serum contains novel trypanocidal protein. J. Euk. Microbiol. **41**:95-103.

Reichl, D. 1990. Lipoproteins of human peripheral lymph. Eur. Heart J. **11**:230-236.

Reinitz, D.M., B.D. Alzenstein, and J.M. Mansfield. 1992. Variable and conserved structural elements of trypanosome variant glycoproteins. Mol. Biochem. Parasitol. **51**: 119-132.

Salmon, D., M. Geuskens, F. Hanocq, J. Hanocq-Quertier, D. Nolan, L. Ruben, and E. Pays. 1994. A novel heterodimeric transferrin receptor encoded by a pair of VSG expression site-associated genes in *T. brucei*. Cell **78**: 75-86.

Schell, D., N.K. Borowy, and P. Overath. 1991b. Transferrin is a growth factor for the bloodstream form of *Trypanosoma brucei*. Parasitol. Res. **77**: 558-560.

Seed, J.R. and S.J. Black, 1999. A revised arithmetic model of long slender to short stumpy transformation in the African trypanosomes. J. Parasitol. **85**:850-854.

Shakibaei, M., M. Milaninezhad, and H.J. Risse. 1994. Immunoelectron microscopic studies on the specific adhesion of *Trypanosoma congolense* to cultured vascular endothelial cells. J. Struct. Biol. **112**:125-135.

Ssenyonga G.S., and K.M. Adams. 1975. The number and morphology of trypanosomes in the blood and lymph of rats infected with *Trypanosoma brucei* and *T. congolense*. Parasitology **70**:255-261.

Sternberg, J.M., N.M. Njogu, C.W. Gickhuki, and J.M. Ndungu. 1998. Nitric oxide production in vervet monkeys (*Ceropithecus aethiops*) infected with *Trypanosoma brucei*. Parasite Immunol. **20**:395-397.

Steverding, D., Y.D. Stierhof, H. Fuchs, R. Tauber, and P. Overath. 1995 Transferrin-binding protein complex is the receptor for transferrin uptake in *Trypanosoma brucei*. J. Cell Biol. **131**: 1173-1182.

Stoppini, l., P.A. Buchs, R. Brun, D. Muller, S. Duport, L. Parisi and T. Seebeck. 2000. Infection of organotypic slice cultures from rat central nervous tissue with *Trypanosoma brucei brucei*. Int. J. Med. Microbiol. **290**:105-113.

Turner, C.M., J.D. Barry, I. Maudlin, and K. Vickerman. 1988. An estimate of the size of the metacyclic variable antigen repertoire of *Trypanosoma brucei rhodesiense*. Parasitology, **97**: 269-276.

Vaidya, T., M. Bakhiet, K. L. Hill, T. Olsson, K. Kristensson, and J. E. Donelson. 1997. The Gene For a T Lymphocyte Triggering Factor From African Trypanosomes. Journal of Experimental Medicine **186**:433.

Wang, Q, N.B. Murphy, and S.J. Black. 1999. Infection-associated decline of Cape buffalo blood catalase augments serum trypanocidal activity. Inf. Imm. **67**, 2797-803.

Wang J., A. Van Praagh, E. Hamilton, Q. Wang, B. Zou, M. Muranjan, N.B. Murphy, and S.J. Black. 2001. Serum xanthine oxidase and control of African trypanosome infections in Cape buffalo. AntiOxidants and Redox Signaling. *In press*.

Ziegelbauer, K., and P. Overath. 1992. Identification of invariant surface glycoproteins in the bloodstream stage of *Trypanosoma brucei*. J. Biol. Chem. **267**: 10791-10796.

EPILOGUE

The opening chapter in this volume expresses Dr. Molyneux's frustration with the scientific enterprise as it applies to controlling the African trypanosomiases. This frustration reflects: (i) our failure to sustain the traditional methods of vector and disease control while pressing forward with the development of new technologies, (ii) our failure to capitalize on scientific findings through development of better vector control strategies, appropriate specific and affordable diagnostic tests, novel chemotherapeutic agents, and ultimately effective immunization protocols, (iii) our habit of justifying requests for further research funding by stressing the need for the above, and (iv) our failure to learn from past mistakes in tsetse control. Some of these concerns are also raised in the essays of Buscher (p51-64) and Seed and Boykin (p65-78). Dr. Molyneux's essay calls us to ponder whether our scientific enterprise is a sham, an uncoordinated but never-the-less honest attempt to meet the challenges posed by the African trypanosomes, or simply a data collecting exercise that is indifferent to individual suffering. The ensuing chapters in the volume address many of the issues raised by Dr. Molyneux and provide the reader with material on which to evaluate these issues.

We share Dr. Molyneux's concern for the abysmal conditions under which many in Africa live, and the apparent absence of a well-thought out plan through which the public health crisis that is enveloping the African continent can be addressed. Trypansomiasis is but one problem among many. We remain hopeful that our leaders in the political arena will make a firm commitment to rid the world of disease, poverty and ignorance, and act on it. We are grateful that private foundations are stepping into the breach to support research aimed at achieving this desired future, and are excited by the tools that modern parasitologists are developing to help control disease.

We consider that PAATIS (Gilbert *et al.,* p11-24) and the future trypanosomiasis risk assessment program (McDermott *et al.,* p25-38) discussed in this volume are important steps towards developing an appropriate long-term plan for trypanosomiasis control. We also share the excitement of investigators who are pushing forward with individual plans to address the problem of trypanosomiasis, which include: the engineering of trypanocidal tsetse flies (Aksoy, p39-50), better diagnostic systems (Buscher, p51-64), the development and identification of novel chemotherapeutic agents by combinatorial chemistry and selective screening (Seed and

Boykin, p65-78), the exploitation of quorum-sensing processes to limit trypanosome population growth (Murphy and Olijhoek, p113-126), and the development of a vaccine that targets the trypanosome flagellar pocket (Black *et al.*, p159-174), which is underpinned by the elucidation of immune mechanisms in trypanosome infected animals (Mansfield *et al.*, p79-96; Naessens *et al.*, p97-112), the dissection of trypanosome endocytic mechanisms (Nolan *et al.*, p127-142), and the full characterization of the trypanosome genome (Donelson, p143-158).

Not all of the exciting research that is being carried out on trypanosomes is included in the volume. Glaring omissions include the on-going development of tsetse attractants and tsetse-trap technologies, the investigation of the molecular basis for human resistance to infection with *T. b. brucei, T. congolense* and *T. vivax*, which may lead to the engineering of Nagana-resistant cattle, and the characterization of metabolic and transport processes that are critical for trypanosome survival and the focus of targeted drug design. Circumstances beyond our control resulted in these omissions, which can be addressed in future texts.

We hope that you have enjoyed the volume and share our sense of pleasure in the achievements of our field.

Samuel J. Black, John R. Seed, 26 June 2001

INDEX